energy and the environment

Theodore L. Brown
University of Illinois

Charles E. Merrill Publishing Company
A Bell & Howell Company
Columbus, Ohio

ISBN: 0-675-09815-7
Library of Congress Catalog Number: 76-161673

1 2 3 4 5 6 7 8 9 10—79 78 77 76 75 74 73 72 71

Printed in the United States of America

prologue

We who are living in the twentieth century and will be living in the twenty-first occupy a special position in human history. We must participate in the transition, if there is to be one, from a human society expanding and developing to the limits of the planet to one living within the means imposed by its closed, limited resources. To negotiate successfully this critical period in history, mankind must learn how to control and stop the growth of population; it must learn how to avoid major war; it must accept the fact that there is a limit on the total rate at which energy can be generated on the planet.

This book is, in part, about the last of these, the limits on man's consumption of energy. It is also about more immediately urgent matters relating to energy which must be dealt with if world society and our nation are to continue to advance materially.

The broad question of how man's activities might affect the global climate is of great importance for the future of the human race. Unfortunately, not much of any value has been written about it for the general audience. For example, the only material which treats in any detail the effects of atmospheric carbon dioxide on climate is in the research literature. The "greenhouse effect" of carbon dioxide gets a line or paragraph in almost every book and article dealing with environment or the population explosion. Anyone wishing to formulate a responsible and informed opinion on the matter would, however, find intelligible sources difficult to locate.

I have attempted in this book to describe how man's generation and consumption of energy can affect the quality and character of his natural surroundings in the present and for the future. I have been as detailed as I felt I could be, consistent with clarity and readability. What I have written is based upon a thorough reading of the recent literature and other information sources. On occasion I have consulted research scientists and others working in relevant areas to obtain opinions about a particular work or idea. I must assume full responsibility, however, for the views expressed in this book not specifically attributed to others.

If you happen to be a professional climatologist, meteorologist, limnologist, or oceanographer, you will soon recognize that I am an amateur in your domain of special knowledge. But if my lack of professionalism in these areas should occasion a lapse here and there with respect to details, I trust that what I have put together will not be found wanting in substantive matters. And if lack of detailed professional expertise must count as a debit, perhaps freedom from biases rooted in professional activity will serve as a counterbalancing credit. In any case, nearly twenty years of experience as a teacher and research chemist have stood me in good stead in sifting through an enormous mass of data and opinion.

After much deliberation I have decided to forego a strict use of metric units in favor of a rather loose mixture which optimizes the accessibility of the material for the reader who lacks an extensive science background. I realize that in following this plan I may incur the ire of some science teachers. Nevertheless, the obligation of the writer is to communicate with his intended audience, and I feel sure that use of non-metric units and the Fahrenheit temperature scale will more forcefully bring home the points I want to make.

Although to a non-scientist there may seem to be a great many numbers in the book, there are far fewer than there might have been. If the exposition seems cryptic in a few places, the message will, I hope, nevertheless come through. And the message is important. The continued burning of fossil fuels at the rates now projected into the next century would result in a very sizable increase in the average global temperature, with disastrous consequences in terms of climatic change. This temperature increase would arise because of carbon dioxide added to the atmosphere in the combustion of these fuels. In addition to this contribution to an increase in global temperature, a steadily increasing rate of energy consumption by man, through use of fuels of all kinds, will eventually exert an additional warming influence on the global climate. Before the global effects of man-made energy dissipation are evident, however, problems in disposing of heat on more local levels will have become acute. In the United States difficulties in finding suitable locations for new steam electric plants are already being experienced in some regions. Man-made contributions to the heat budget of a heavily populated region such as the eastern seaboard may be expected to produce serious climatic changes during the next thirty years. It is not too soon to recognize the impending difficulties and begin the establishment of a national policy for energy. This policy will necessarily include allocation of total energy consumption on a regional basis and establishment of controls on the rates of growth of various components of the energy market, such as the electric utility industry.

I hope that the survey of present and future prospects for energy use presented herein will convince the reader that there are limitations on the use of energy imposed by the current state of technology and the need to preserve the environment. At the very least, the numbers put forth in the book may be of benefit to some.

It may be helpful to briefly describe the organization of the book. The first three chapters are devoted to a brief introduction to the atmosphere, the earth's radiation budget, the development of atmospheric temperature and climate. There is much interesting and quite comprehensible physics in it and I think you will find it enjoyable reading. Use is made of this background in Chapter 4, in which the "greenhouse effect" and the influence of atmospheric carbon dioxide on climate are thoroughly explicated. Chapter 5 is concerned with particulate matter as it affects climate. I have included a discussion of the possible climatic effects of SST flights in this chapter.

In Chapter 6 the question of energy consumption on global and regional scales is dealt with. Chapter 7 treats the more localized consequences of heat generation, particularly in terms of effects on water bodies. In Chapter 8 I have considered a few important but typical examples of the effects of power plant installations on the temperatures of water bodies used in their cooling.

In Chapter 9 I have summarized the findings of earlier chapters and put them into the perspective of contemporary thought regarding climatic change. The prospects of calamitous alterations in climate resulting from an uncontrolled increase in the consumption of fossil fuels, and from an excessively high total rate of energy generation, are real. I have set forth a few suggestions about how we might formulate a new kind of conservatism with respect to energy use, on both national and international levels.

Although the subject matter of this book does not make for light-hearted writing (or reading), I must say that I have enjoyed writing it. Mankind's present and future problems with energy are, after all, products of his very considerable success in adapting his surroundings to his purposes as he has seen them. There is no compelling reason to suppose that he will fail to develop controls on the uses of energy so as to preserve the quality of the environment. The actions required to establish these controls will not be taken, however, unless there is widespread appreciation of their necessity. The hope that this little book might foster a broader and deeper understanding of the issues involved has stimulated my efforts. My family have helped with their advice and by showing considerable forbearance with my preoccupation while the book was in progress. I am grateful also to Mr. Ray Sweany, who read and criticized an early draft version.

T. L. Brown

Champaign, Illinois

table of contents

1 this most excellent canopy, the air

Suppose you are lying out on the beach or in an open field on one of those spectacularly clear, bright days that happen occasionally. You look up into the zenith, the blue infinity of sky. What is it—what is the sky, the impalpable blue something, a medium invisible, yet possessing endless variety and awesome power?

The bright, even blueness of the sky is evidence of the presence above us of a thick, enveloping cover of atmosphere. Without this protective cover, a human standing on the surface of the earth would look out into a universe black everywhere except for the light from distant stars, the reflected light of the moon, and the blaze of the sun. The atmosphere filters the radiations and all forms of matter which impinge upon earth from outer space and removes or alters most of them before they reach the surface. Thus life on earth is protected from excessive amounts of harmful high energy radiation. The rays to which we refer as **visible light** pass through the atmosphere with only a small diminution in intensity. In the process, however, they may be scattered many times by molecules of the atmosphere. The efficiency of scattering is not the same for all the various colors of the visible light spectrum. The overall effect is a blue color in the sky because blue light is the most efficiently scattered part of the visible spectrum. The scattering of light diffuses the sun's rays and provides an evenly distributed daylight over the entire sky. If it were not for this effect, the earth would be an eerie place of intense brightness and deep shadow.

We breathe the air, sail kites, fly airplanes and blow smoke in it. It waxes hot and cold, propels sailboats, howls through abandoned castles, dries the wash hung on the line. For nearly all of us, contact with the atmosphere is confined to occurrences at the bottom of a kind of sea. We might think of the atmosphere as an ocean of gas, covering the surface of the globe. The gas stays on the surface instead of drifting out into space because it is influenced by the gravitational attraction for the planet experienced by each and every molecule of the atmosphere. These molecules are in constant motion, colliding with one another and with solid objects. The motions of the molecules, however, when averaged over a period of time, reflect the influence of the gravitational force.

Suppose we were out in space, a few hundred miles above earth, and could drop in a straight line down to the surface. At our starting place there would be an extremely few stray molecules, and a thin scattering of matter from outer space. As we began the downward plunge we would encounter increasing numbers of gas molecules in each unit of volume. This number of molecules per unit volume would steadily increase as we approached the surface. In other words, the density of the atmosphere increases with closeness to the surface. The molecules of the gas further out press in, so to speak, on the ones below them. The result is quite dramatic; 95% of the total atmosphere is in the layer from the surface to a height of 12 miles, even though we consider the atmosphere to extend more than a hundred miles into space.

Meteorologists, who study the earth's atmosphere, have divided the atmosphere into layers for the purpose of distinguishing certain characteristics which change dramatically with height. These layers are depicted in Figure 1-1. The lowest layer, in which all but a few of us are destined to remain for our entire lives, is called the **troposphere.** The troposphere is where most of the action is, or at least the action we can observe. Thunder and lightning, rain and rainbows, hail, snow and spectacular sunsets—all derive from processes occurring in the layer of atmosphere which extends to about 11 miles in the tropics and 5 miles in the polar regions.

In moving up from the surface through the troposphere, the temperature decreases steadily, at a rate of about 3.6°F per

1000 feet (6.5°C per kilometer). This rate of temperature decreases with increasing altitude is called the **lapse rate.** The **tropopause,** which is the boundary between the troposphere and the stratosphere, marks the height at which the temperature reaches a minimum. At this height the atmosphere is very cold. Many people will have had the experience of flying in a modern jet passenger plane, and hearing the pilot say, "Ladies and gentlemen, we are now flying at an altitude of 31,000 feet, which is just about 6 miles, and passing over Columbus, Ohio. The temperature on the ground in Columbus is 87°; the temperature outside our cabin windows up here is 45° . . . below." "Oh, my!" says the little old lady, sipping her second martini. Yes, it is very cold out there; and since 6 miles up is not far from the tropopause, that is about as cold as it gets.

The lapse rate we have been discussing also applies in about the same degree for increase in altitude above sea level on land. The variation in temperature with altitude is extremely important in determining climate. Thus, at the peak of Mount Everest, at 29,000 feet, it is extremely cold. The Altiplano plateau high in the Andes in Peru has a cold, arid climate, whereas on the coastal plain not far away it is hot and moist.

Directly above the troposphere is the stratosphere, extending to a height of about 30 miles (50 kilometers). It was discovered, in a manner of speaking, at about the turn of the century by a French meteorologist with the impressive name of Leon Phillipe Teisserenc de Bort. Before Teisserenc de Bort's work, it was commonly believed that the temperature of the atmosphere just steadily decreased until it reached absolute zero,* and that beyond this there was just the void of outer space. He sent up temperature-measuring devices in balloons, and found that, at an altitude of about 7 miles, the temperature stopped falling, and remained constant with further increase in altitude. More modern work, involving measurements at higher altitudes, has shown that the stratospheric temperature does not remain constant, but actually begins to increase above about 12 miles. At an altitude of about 20 miles the temperature is not much lower than on the surface of the earth.

Although the temperature is right, a man at this altitude

---

* Absolute zero is the lowest possible temperature, at which all heat motion of atoms and molecules ceases. It corresponds to −459°F.

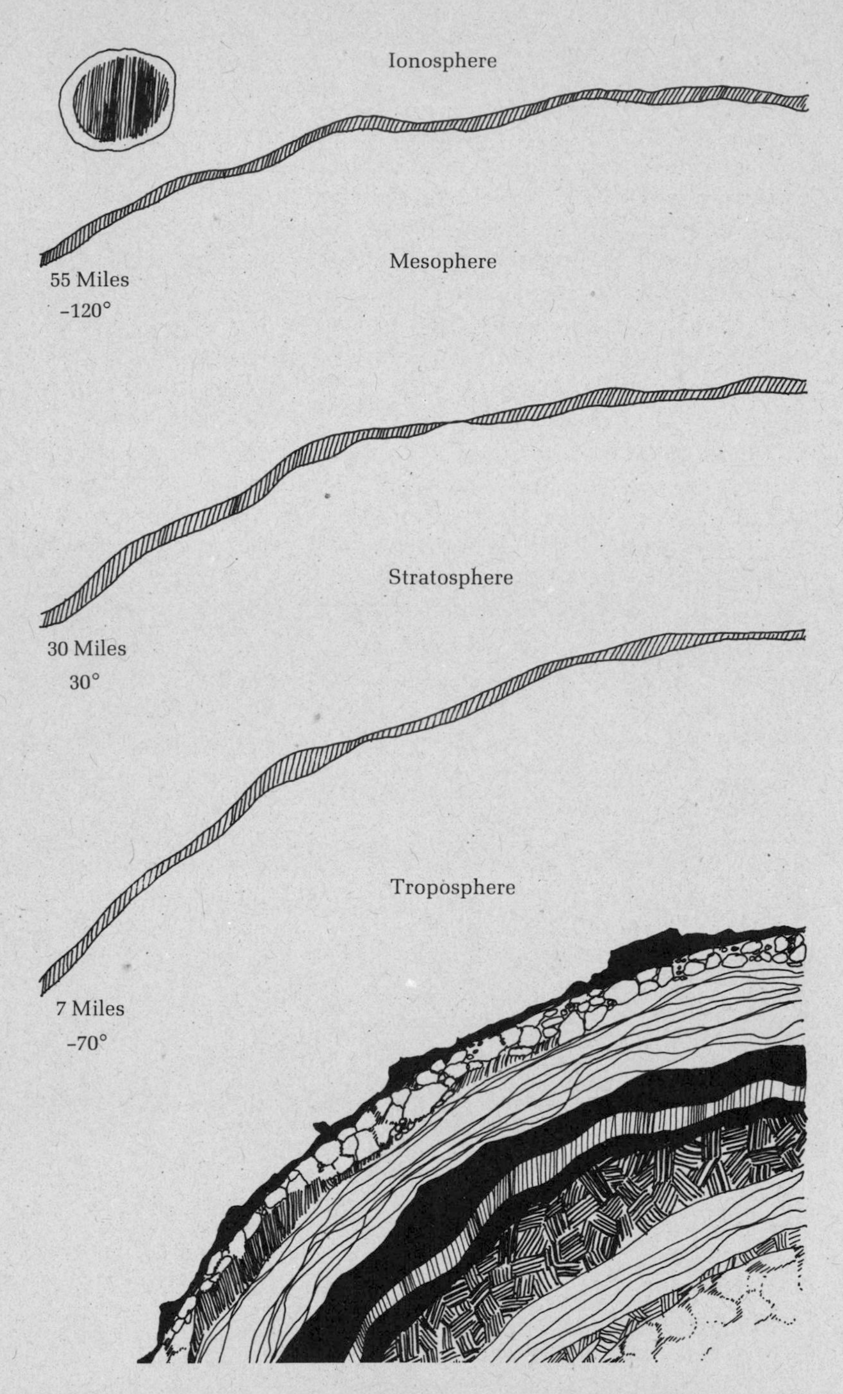

FIGURE 1–1. *The regions of the atmosphere. The altitudes and temperatures (°F) are average values.*

would not find the upper stratosphere a hospitable place. Because of the drop in pressure with increasing altitude, the atmosphere at 20 miles is extremely thin; the pressure is only one hundredth of that on the surface. Our man would essentially be in a vacuum. In addition, much of the protective effect of the atmosphere is below this height. Intense, damaging ultraviolet radiation from the sun would beat with undiminished vigor. In fact, the reason for the increase in temperature in the stratosphere in the range from 12 to 20 miles is the absorption of higher energy ultraviolet radiation from the sun.

The most important substance in this absorption process is ozone, a molecule consisting of three atoms of oxygen, which chemists write as O_3. Ozone is not a very stable molecule; it is formed in the stratosphere by the action of high energy solar radiation on ordinary oxygen molecules, which consist of two atoms of oxygen bound together, written as O_2.The high energy radiation is absorbed and causes a dissociation, or breaking apart, of the oxygen molecule, to form two oxygen atoms. The chemical equation for this simple process is

$$\text{light} + O_2 \longrightarrow O + O$$

The oxygen atoms produced in this process are very reactive. When they collide under the appropriate circumstances with an ordinary oxygen molecule an ozone molecule is formed:

$$O + O_2 \longrightarrow O_3 + \text{heat}$$

The ozone thus formed may eventually decompose into oxygen molecules by colliding with another free oxygen atom:

$$O_3 + O \longrightarrow O_2 + O_2$$

or they may absorb ultraviolet rays from the sun. If energy from the sun is absorbed, a reversal of the reaction which formed the ozone in the first place occurs:

$$\text{light} + O_3 \longrightarrow O_2 + O + \text{heat}$$

Eventually the oxygen atoms which are formed in these processes recombine to form stable O_2 molecules, and heat is given off. Thus, there is an "ozone machine" in the stratosphere which takes out a band of high energy, ultraviolet rays from the solar radiation and converts the energy of these radiations into heat which warms the stratosphere. The processes are dynamic, so there is a continual chemical cycling going on. Some reactions are more important in the daylight hours, others at night. The very fortunate result of all this is that the harmful ultraviolet rays are filtered out. An additional consequence is that there is a large quantity of ozone in the stratosphere. In terms of the total atmospheric composition it is not great, only about 10 parts per million at most (which means simply that there are about 10 ozone molecules in a sample of the atmosphere which contains a million total molecules). By contrast, however, the concentration of ozone at the surface is much lower, typically about 2 to 5 parts per hundred million.[1] Aside from its role as a radiation filter, ozone plays a minor part in determining the heat balance of the earth, and we shall refer to it again in a later chapter.

Above the stratosphere, extending from an altitude of 30 miles to about 55 miles, lies the **mesosphere,** an intermediate region in which the atmosphere grows continuously thinner and colder. At the top of the mesosphere the temperature is only about –120°F, and the pressure is only about one ten-thousandth as great as on the surface. In this thin, frigid region the sun's rays work a simple but powerful chemistry on the atmosphere. This chemistry is even more extensively seen in the vast region above the mesosphere, called the **ionosphere.** Although the density of molecules in the outer reaches of the atmosphere is very low, the powerful force of the sun's high energy radiations produces extensive effects. The highest energy ultraviolet rays, x-rays and cosmic rays emanating from the sun impinge on this outer bastion of the earthly envelope and produce not merely decomposition of molecules into atoms, but much more extensive changes as well. These various forms of radiation are capable of stripping electrons from atoms and molecules to form positive ions and free electrons in space. In the upper regions of the ionosphere the ions and electrons do not recombine quickly because the particle concentrations are extremely low. This means that a long time may pass before a particular ion or electron encounters a particle of opposite sign with which it may combine. Thus, the ionic character of the ionosphere is preserved even at night in the absence of solar radiation, although the concentrations of ions and electrons gradually diminish.

The ionosphere is an extremely interesting part of the atmosphere, and has been extensively studied since its existence was demonstrated by experiments carried out in the 1920s. It refracts (bends) radio waves, and is responsible for the fact that radio broadcasts can be received at places far over the horizon, even halfway around the globe, from the transmitter. It does not play an important part, however, in determining the earth's heat balance, so we will not consider it further in this book. Many other fascinating atmospheric phenomena such as auroras, airglow, the Van Allen radiation belts and others are not germane to our immediate concern and will not be considered.

The atmosphere consists of a number of more or less permanent components, listed in Table 1. The major constituents are nitrogen and oxygen. Argon is an entirely inert gas, and does not participate in chemical reactions at the earth's surface, although it may be ionized in the ionosphere. Carbon dioxide is an important ingredient because it is necessary for plant life. It is also important in establishment of the earth's heat balance.

TABLE 1. *The composition of clean, dry air near sea level.*

Component	*Content* (Volume percent)
Nitrogen	78.084
Oxygen	20.9476
Argon	0.934
Carbon Dioxide	0.0318
Neon	0.00182
Helium	0.00052
Krypton	0.00011
Xenon	0.000009
Hydrogen	0.00005
Methane	0.0002
Nitrous Oxide	0.00005
Ozone	0.000004

The remaining components of the atmosphere are all present to trifling extents and are not important for our purposes.

In addition to the gases listed in Table 1, a number of others are present in the lower atmosphere, largely as a result of their generation by man. These include sulfur dioxide, nitrogen dioxide, ammonia and carbon monoxide, and ozone in concentrations in excess of that given in Table 1. These pollutant gases are deleterious in many ways when present in locally high concentrations. Thoroughly dispersed in the entire atmosphere, however, they are not evidently harmful. Their lifetime in the atmospheric system is limited because they are converted to some other compound and "scrubbed out" through rainfall or in the atmosphere-ocean interaction.

The composition of the atmosphere as listed in Table 1 is given for dry air and omits water vapor, a most important ingredient. The water vapor level of air at sea level may vary from as high as 20 parts per thousand in the tropics to less than one part per thousand in dry polar regions. The water cycle, termed the hydrologic cycle, involves interaction of the atmosphere with the surface waters of the oceans, lakes, and other water bodies; evaporation from plants; rainfall and snow; and large scale movements of air masses. This brings us to the next element in the development of requisite background material, the earth's radiation budget, and the link between this and climate.

2 here comes the sun

Radiation from the sun is absolutely necessary for sustaining life on earth. If the sun were suddenly to go out like a snuffed candle, we would continue to spin about in space on a dark planet growing colder day by day. Soon the oceans would grow cold enough to freeze, the world-wide atmospheric circulation system would be disrupted; fierce storms would tear across the planet. Eventually, after all life activity had ceased, the major components of the atmosphere, nitrogen, oxygen and argon, would liquify in the numbing cold. Still later, even these substances would solidify, leaving the earth with a thin atmosphere of helium and hydrogen vapor over the frozen surface.

The energy received from the sun powers a complex set of processes which distribute energy, in the form of heat, over the planet. This redistribution provides the reasonably even climatic conditions which make life possible over a large fraction of the earth's surface. In order to understand how these processes work, and therefore to have some understanding of climate, we must look in some detail into the earth's radiation balance.

Figure 2-1 shows how the earth intercepts a bundle of rays from the sun. At any given instant, some of these rays are falling on an area of the earth which is perpendicular to the direction of the rays, while other areas of the earth slant away from this direction. As a result, a unit area in the polar regions of the earth does not receive as much solar radiation, averaged

FIGURE 2–1. *Solar radiation intercepted by earth. The radiation falling in one minute upon a unit cross-sectional area, as shown in the insert, is the solar constant. At any particular time most of the earth is not perpendicular to the sun's rays. A unit area parallel to the earth's surface receives much less energy per unit time, when averaged over the entire year, than is represented by the solar constant. The amount received varies with latitude and is least at the poles.*

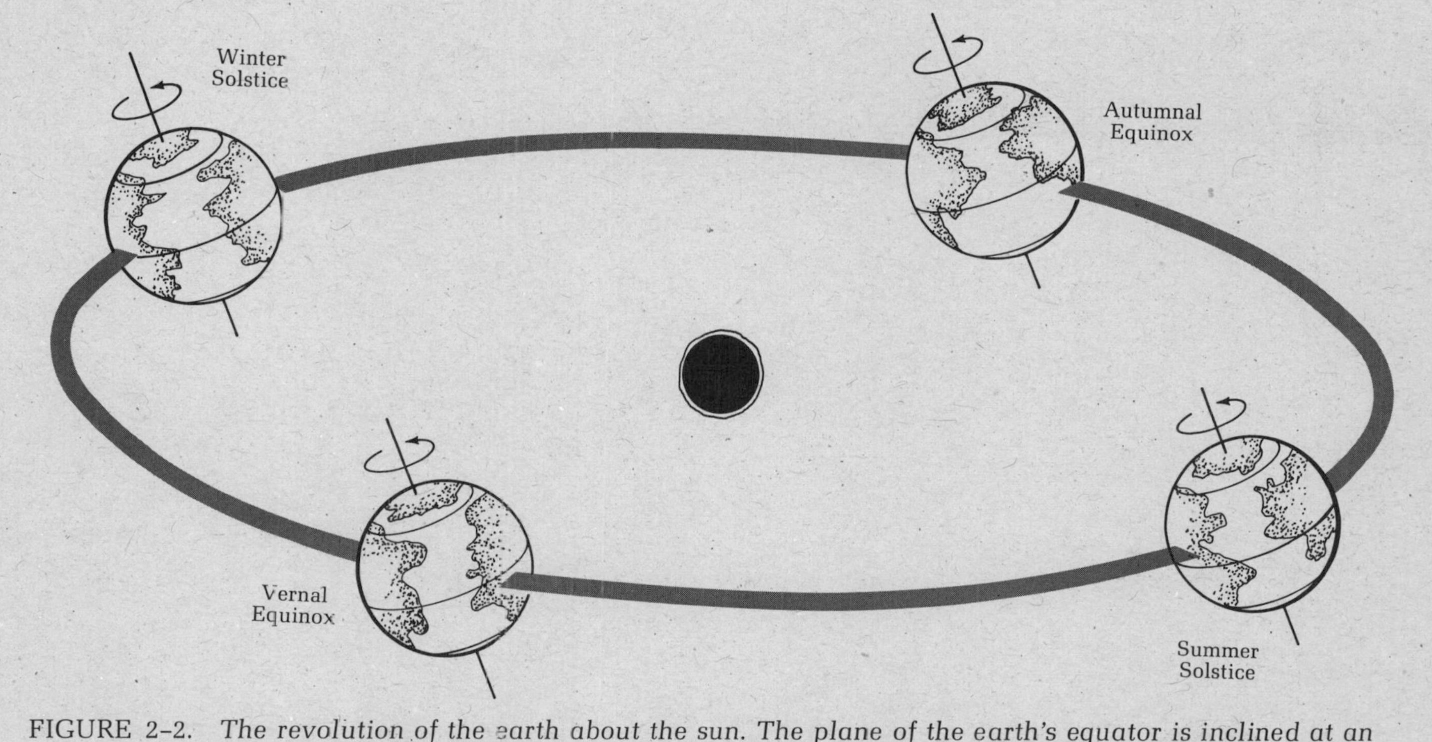

FIGURE 2–2. *The revolution of the earth about the sun. The plane of the earth's equator is inclined at an angle of 23½ degrees from the plane of the earth's orbit about the sun. It is this inclination which gives rise to the seasons.*

over the year, as an equal area in the equatorial zone. Since the earth's axis is tilted with respect to the plane in which it rotates about the sun, Figure 2-2, the average amount of solar radiation which falls in a certain area in a day varies with the time of year for any particular area. Thus, as we all know, it is winter in the southern hemisphere at the time when it is summer in the northern hemisphere.

The **solar constant** is defined as the quantity of radiant energy which falls in one minute on a unit area of one square centimeter,* exposed perpendicularly to the sun's rays. This energy is measured in calories.† S. P. Langley (1834–1906), an American physicist who was associated with the Smithsonian Institution for many years, was among the first to attempt measurement of the solar constant. In his honor, a unit equivalent to 1 calorie per cm^2 is termed the **langley**.

One of the first things we need to ask about this solar constant is whether it is really constant. Astronomers who study the sun know that it is far from a steadily burning ball of atomic fire. Huge solar flares lasting from one to ten minutes and extending over millions of square miles of the sun's surface send out enormous torrents of high energy particles and radiation. Sunspots, which seem to be regions of especially intense activity, also result in changes in the radiation reaching earth from the sun. Sunspot activity has been monitored by astronomers for many years and has been observed to vary periodically, primarily with an 11-year cycle.

Attempts have been made over the years to measure the solar constant. It is important to know whether it does indeed vary, by how much, and whether the variation is periodic in character. But the solar constant is exceedingly difficult to evaluate. Recall that it is a measure of the radiation falling on a unit area perpendicular to the sun's rays. Ideally we would measure this quantity *before* the sun's radiation penetrates the earth's atmosphere, where a number of processes may occur to remove some portions of the radiation. In the past the best that could be done in the way of systematic long-term studies was

---

*There are 2.54 centimeters in one inch, so 6.5 square centimeters (cm^2) are equivalent to one square inch.

†To be precise we should say a gram-calorie. A gram-calorie is the energy required to raise the temperature of one gram of water by one degree centigrade, or 1.8 degrees Fahrenheit.

to set up measurement on the highest accessible mountains, which included Mount Wilson and Table Mountain in California. These measurements, carried out for many years under the sponsorship of the Smithsonian Institution, showed considerable variations in the solar constant. Much of the apparent variation, however, has since been found to be attributable to changes in the earth's atmosphere. Accurate corrections for these variations could not be made until recently.

The solar constant can now be measured with good accuracy from the "top of the atmosphere" using orbiting meteorological satellites. The first accurate measurements are now beginning to come in.[1] They indicate that the solar constant is 1.99 langley per minute, with an accuracy of about 1½%. More extensive measurements will be needed to reveal whether this value changes materially with time, but the present indications are that the solar constant is really, to a good approximation, a constant.

This is a very important assumption when one is discussing climate. The sun's radiation is the sole source of energy for atmospheric processes and for maintenance of the earth's climate. A change of even a few per cent in the solar constant for a protracted period could produce important climatic changes on earth. Many a rather handsome theory of climatic change during the earth's recent history has been constructed on the notion of a variation in solar constant. There is no conclusive evidence one way or the other, however, regarding the importance of solar variation in determining long term, massive climatic changes. Attempts at correlating sunspot activity with extent of glaciation or other indices of climatic change in more recent times have not been notably successful. We may provisionally say, then, that (1) the solar constant is really quite constant and (2) sunspot activity, which varies with an 11-year period, causes variation in the high energy ultraviolet and x-radiation reaching the earth, but this does not seem to result in significant climatic variation.

To develop the picture of the earth's energy balance further, we need to look into the nature of the energy reaching earth from the sun. First, we recognize that it is radiant energy, or electromagnetic radiation, which travels through space with a speed of 186,000 miles per second. Thus, it takes the radiation from the sun about 8 minutes to make the trip to earth. This speed is much greater than the speed at which even the smallest material particles can travel. For example, the sun also throws off enormous quantities of sub-atomic charged particles such as electrons and protons. Bursts of these particles, which possess tremendous energies, arrive at earth some 15–30 hours after leaving the sun during a solar flare, and produce the spectacular luminous auroras seen in the polar regions.

Thus electromagnetic radiation is a non-material, periodic energy phenomenon which moves through space at the characteristic speed of 186,000 miles per second. The radiation is characterized by a frequency; it is as though there were waves of energy moving through space. Suppose you could stand to one side and watch these waves go by (at 186,000 miles per second!). The number of energy maxima which would pass your observation point in one second is the frequency of the radiation. This model situation, scientifically very gauche, but still useful, is illustrated in Figure 2–3. Note that if the frequency is higher, the separation between maxima is shorter. This distance, termed the wavelength, is frequently used to describe radiation. It is useful to remember that the higher frequency radiations possess shorter wavelengths.

All the various forms of electromagnetic radiation: radio broadcast signals, television broadcast signals, the emanations from infrared heat lamps, light bulbs, x-ray tubes, the sun, red-hot charcoal, and many, many others are basically the same phenomenon. They differ only in the frequency of the radiation, and thus in wavelength, but travel through space at the same speed, 186,000 miles per second. They obviously differ greatly, though, in their effects on substances with which they come in contact. We may say, in a rather imprecise way, that the energy content of a given radiation increases with increasing frequency. Thus, the low frequency radiations such as broadcast waves do not affect matter, and, in fact, pass through materials which are not electrically conductive. Higher energy (higher frequency) radiation such as infrared or visible light are absorbed or reflected by matter. The absorption may result in a heating of the material as with infrared radiation, or a chemical change (fading of colors, sunburning) in the case of the higher energy visible and near-ultraviolet radiation. Radiation in the ultraviolet and x-ray region is of extremely high energy and very short wavelength. It has the capacity to effect profound chemical change. We saw in the previous chapter how the upper regions of the atmosphere act to filter the ex-

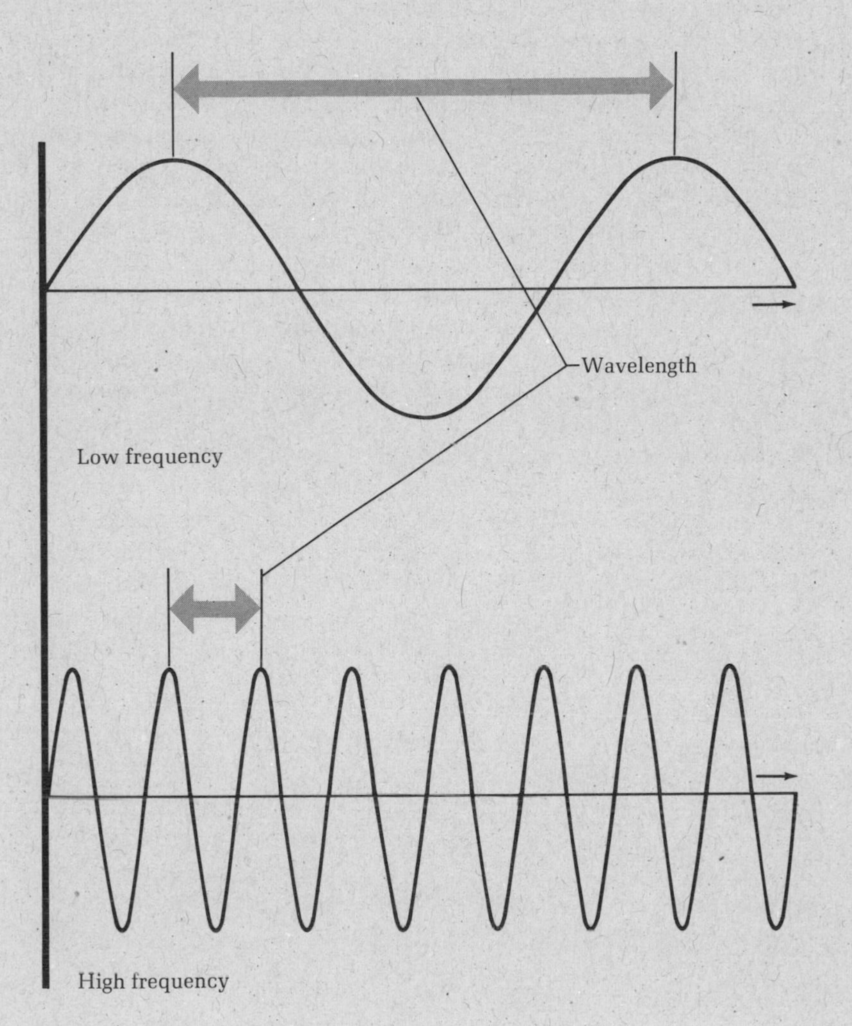

FIGURE 2-3. *Relationship between frequency and wavelength of electromagnetic radiation.*

tremely high energy radiations coming from the sun before they reach the surface. This filtering occurs as the molecules of the atmosphere absorb the high energy radiation, with a dissociation of the molecule, ionization (tearing off of an electron), or some other high energy process. The lower energy radiation passes through the upper reaches of the atmosphere because there are no processes which might occur there which are low enough in energy to use it.

The sun's radiation spans a very wide range of wavelengths. This distribution of wavelengths is termed a spectrum. Figure 2-4 gives the ranges of wavelengths for radiations in the various familiar regions of the spectrum. The unit of wavelength most commonly employed in discussing radiation is the micron, which is one millionth of a meter.* Visible light has a wavelength a few tenths of this little micron. The human eye is capable of discerning radiation which has a wavelength ranging from 0.4 micron to about 0.7 micron. This is a tiny "window" in the vast span of solar radiation wavelengths. It just happens to be the only window in the entire solar radiation which passes with little or no diminution to the earth's surface.

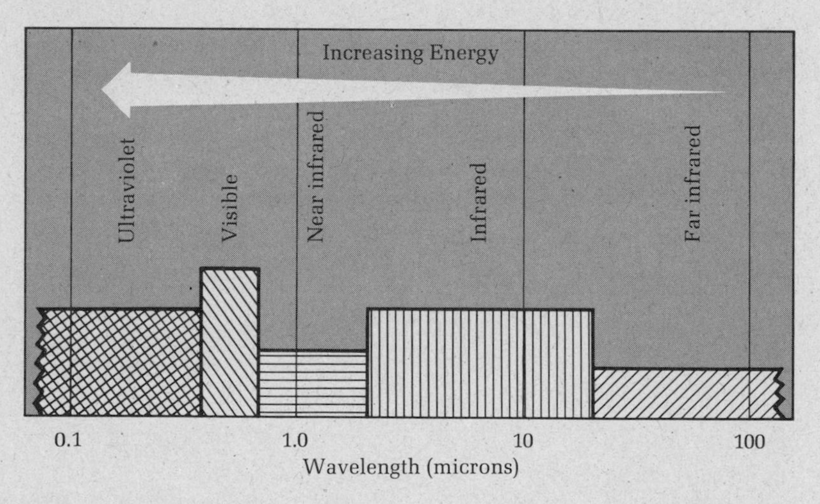

FIGURE 2-4. *Wavelength ranges of the characteristic regions of the spectrum.*

---

*A meter is slightly longer than a yard. Imagine a yardstick divided into one million divisions; each of these divisions is about a micron.

Every object is a source of radiant energy. The radiation arises from motions of the electrons of the atoms and molecules of the substance. These motions represent the heat energy which the object possesses. The hotter it is, the more extensive and higher in frequency the motions may become. We are all familiar with the fact that when an object is heated to a sufficiently high temperature, it glows a dull red. Heated to still higher temperatures it glows brighter red, and may even become "white hot." These variations in apparent color and intensity are a reflection of the way in which the wavelength and quantity of radiant energy emitted by an object vary with temperature. The intensity distribution of radiations at all the various wavelengths which emanate from an object is termed its **emissivity.**

Physicists in the nineteenth century studied the emissivity of objects. They were interested in knowing whether two objects at the same temperature but composed of different substances would have the same emissivity. It turned out that, in general, they don't, although objects which are quite hot are generally quite similar in their emissions. The physicists came to define an ideal emissivity behavior. At a given temperature there is an upper limit to the amount of radiant energy that can be emitted in a unit time from a unit surface of a body. This upper limit is the **black body radiation.** The black body, then, is a perfect emitter. It has another interesting property—it is also a perfect absorber. In other words, radiation which impinges on the surface of a black body is completely absorbed by it. The **absorptivity** is the fractional part of the radiation impinging on a surface which is absorbed by it. For a black body, which absorbs everything, the absorptivity is one. We can now set up a scale to grade objects. The emissivity is the relative ability of an object to emit radiant energy as compared with the ideal black body at the same temperature. The absorptivity is the fraction of incident radiation absorbed by a surface. Both these quantities can vary between 0 and 1 for real, non-ideal objects.

The term "black body" was chosen to emphasize that the object cannot transmit light (as can glass, for example), or reflect light (as, for example, polished silver), but can only emit radiation or absorb it. It suggests something like a lump of coal, or at the very least a solid object which is "black" with respect to some wavelength of radiation. But any collection of matter capable of absorbing radiation without reflecting or transmitting it is a black body. This collection of matter might be solid, liquid or gas.

We can illustrate these ideas with a simple example. Suppose we cut out a square of aluminum foil and blacken one side with a thick, even coating of chimney soot. Now we suspend this foil near a hot object, say a light bulb, as shown in Figure 2-5. If we mount the aluminum so that the blackened side is toward the light, the aluminum heats up very rapidly. This happens because the absorptivity of the blackened side is very high. On the other hand, the emissivity of the shiny metallic side is low. This is an illustration of Kirchoff's Law, which states that the emissivity of a surface for a given wavelength of radiation is exactly equal to its absorptivity. Since the absorptivity of the aluminum surface is low, because it reflects so much of the radiation, its emissivity must then also be low. As it heats up, of course, the black body radiation, mostly from the same blackened side that is absorbing, increases. Eventually the two processes are in balance, and the temperature of the foil remains constant.

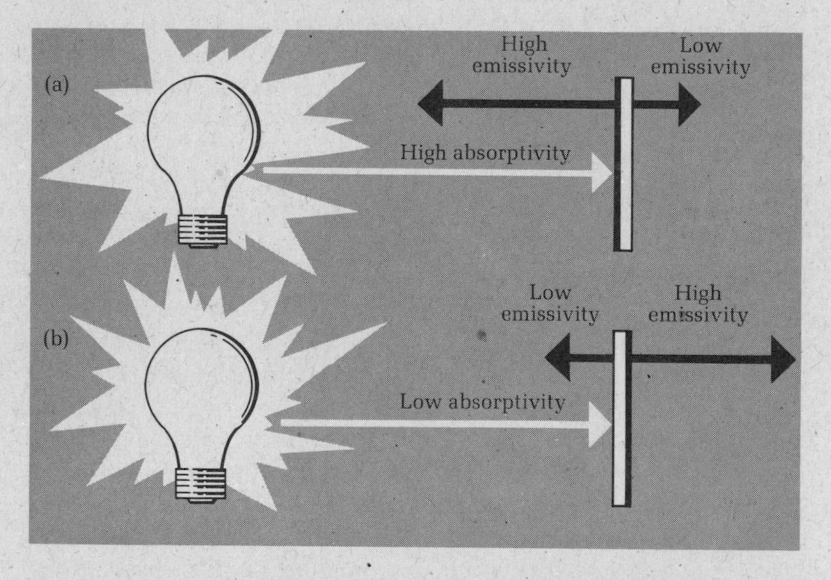

FIGURE 2-5. *Variation in the absorption and emission characteristics of blackened and shiny aluminum surfaces. In (a) the metal attains a higher temperature than in (b).*

Now, for the comparison, let us turn the foil around, so that the shiny side is toward the light bulb. We find that, first of all, the foil heats up much more slowly. The shiny side has a low absorptivity, since much of the light from the bulb which strikes it is reflected away. At the same time, the blackened side is a very good emitter, so from the back side there is an efficient re-radiation of the energy which is absorbed. The aluminum foil in this position, therefore, attains a much lower temperature than in the first instance. By partially blackening the surface exposed to the light bulb, we could establish a steady temperature on the metal foil intermediate between the two extremes we have noted.

Now, what does all this have to do with the earth's radiation balance? Well, if we let the light bulb in our example represent the sun, and the aluminum foil represent mother earth, we have the elements of the situation with which we must deal. The sun is the source; we must therefore ask about its radiation. Is it like that of an ideal black body? If not, what is it like? With respect to the earth, we must ask what sort of absorber of the sun's radiant energy it is: that is, what is the earth's absorptivity toward the wide range of radiations coming from the sun? We must also recognize that although the temperature of earth is not comparable to the sun's it does have a temperature and must therefore radiate energy. Just as in the case of the aluminum foil in our example, there must be an equilibrium between the radiant energy coming into and the radiant energy going out from earth. If the equilibrium did not exist, there would be a net flow of energy in or out, and, consequently, a net steady increase or decrease in temperature for the planet as a whole.

The radiations emitted from a black body as a function of temperature were studied experimentally by the physicists of the last century. They were able to observe the manner in which the quantity of radiant energy varied with wavelength for a black body at some particular temperature, and then the manner in which this distribution changed as the temperature of the body was changed. The results were completely mystifying; no theory of radiation known at the time could account for the experimental findings.

The major breakthrough—indeed one of the major breakthroughs in the history of physical sciences—came in 1900, when Max Planck announced his quantum theory and its application to the black body radiation problem. There is no point in developing a discussion of quantum theory here; we need only note that Planck was able to develop an equation which gave the emissive power of an ideal black body as a function of wavelength for any temperature. This equation exactly reproduced the experimental results. On the basis of Planck's radiation law one can therefore predict the emission vs. wavelength for a black body at any temperature. Three such emission curves are shown in Figure 2-6; one for 11,000°F, the approximate temperature of the outer reaches of the sun, and the other two for 80°F and -100°F. The temperature of earth's atmosphere varies within the range of the latter two temperatures.

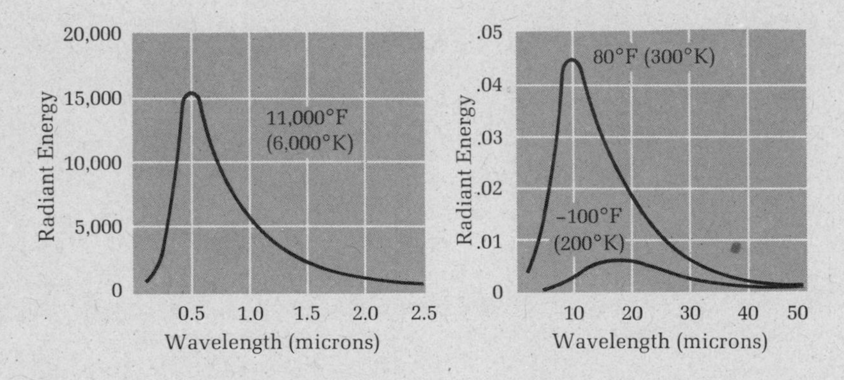

FIGURE 2-6. *Black body emission curves for three different temperatures.*

The quantity plotted along the vertical axis is in units of langley per minute per unit wavelength; this quantity varies with the wavelength of radiation. Note that the scales in the two parts of the figure are vastly different. The 11,000°F object is putting out energy at an incomparably greater rate than the objects at the much lower temperatures. If we tried to graph the data for the 80°F object on the left curve, it would not even appear above the baseline.

The second notable feature of these curves is that they rise to a rather steep maximum. Furthermore, the wavelenth of this maximum varies with temperature. Thus, for the sun's radiation the maximum is 0.48 microns wavelength, in the middle of the visible spectrum (see Figure 2-4), whereas for the earth's radiation the maximum wavelength is in the range from 10 to 14 microns, in the infrared region of the spectrum. It is important in this connection to keep the vertical scales of the two figures in mind. The cooler object does not emit nearly as much radiation at *any* wavelength as the hotter one. Thus, although the sun's radiation has a maximum at 0.48 microns it is still putting out much more radiation in the wavelength interval 10 to 14 microns per unit area than the earth.

The *total* emissive power of an object, as opposed to its emissive power at some particular wavelength, is the sum of all the emissive powers at all the different wavelengths. This total emissive power shows a very sharp dependence on temperature.[2] The two right curves in Figure 2-6 illustrate this very well; changing from 80°F to -100°F causes a huge change in the relative areas under the curves. For our purposes, all we need to recall of this is that a little change in temperature can make a great deal of difference in the amount of energy radiated.

Thus, the total emissive power of the sun is many orders of magnitude larger than that of the earth; the difference is so huge as to be quite unimaginable. But when we consider the consequences of this for the radiation balance of the earth we must remember that the earth is a relatively small object of about 8,000 miles diameter, as compared with its distance from the sun, some 92 million miles. Only a very tiny fraction of the sun's enormous output is intercepted by earth. As the sun's radiation spreads out into space the intensity in a unit cross-sectional area decreases rapidly. If we scale down the radiation curve for the 11,000°F object, to represent just the part which intercepts earth, we must shrink the vertical axis scale, so that the total area under the curve represents a total on the order of 2 langley per minute, the value observed for the solar constant. This scaled-down version of the sun's radiation is shown in Figure 2-7, along with the black body radiation curve for a 40°F object. We see now that the part of the intercepted solar radiation which is important is the range from about 0.2 to 3 microns. These two curves are directly comparable, in that they represent the solar radiation falling during one minute on a unit area perpendicular to the sun's rays, and the emission during one minute from that same area of a black body (e.g., the earth's surface) heated to 40°F (277°K).

Figure 2-7 illustrates something very important. It shows

that the earth receives radiant energy from the sun, mainly concentrated in the visible and near-infrared regions of the spectrum, from 0.2 to 3 microns wavelength. The earth radiates back into space radiation in a wavelength range from about 10 to 14 microns, the infrared region. The energies coming in and going out are in balance. But what kind of balance? What processes must be taking place in the earth's ocean-atmosphere system to establish a climate on earth in the context of this balance? To get at the answers to these questions we must look at the details of the radiation "budget."

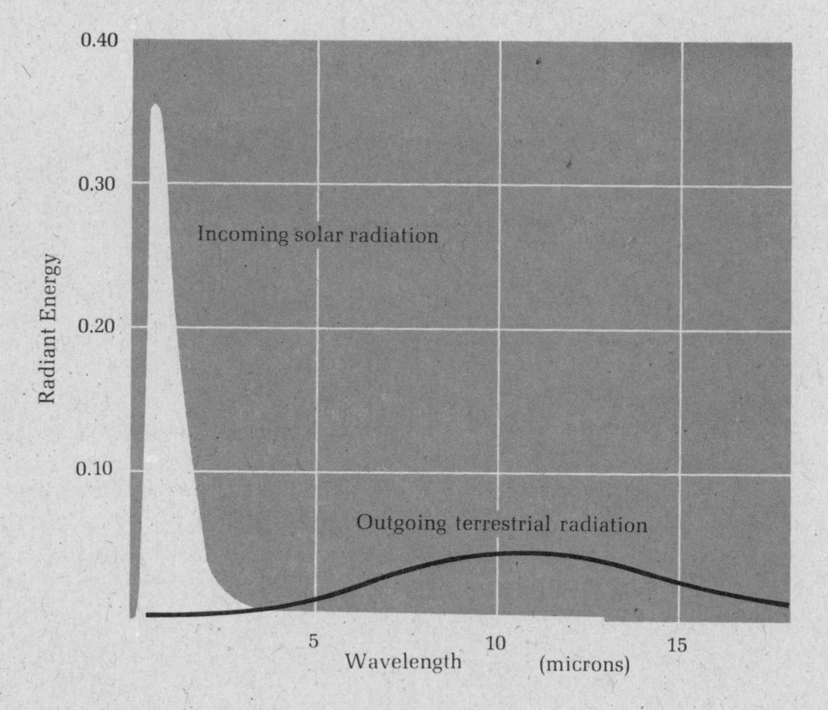

FIGURE 2-7. *Comparison of the wavelength distribution for incoming solar energy and outgoing radiation from earth's surface.*

3 to tell a weather's wind

With the background of the preceding two chapters, we are now in a position to examine in some detail what happens to the solar radiation reaching earth, the processes by which energy is distributed on the planet, and the resulting consequences in terms of climate.

We recall that the solar radiation is concentrated in the wavelength interval from 0.2 to 3 microns. The various fates which might befall this radiation as it passes through the atmosphere are depicted in Figure 3-1. Some may be reflected back into space by clouds or solid particles. Another portion of the radiation is scattered by molecules of the atmosphere. Some light is scattered backwards and thus returned to space; the forward scattering, referred to in the early part of Chapter 1, results in diffusion of light in the atmosphere. A third way in which solar radiation may be returned to space is through reflection from the surface of earth; mainly from water surfaces, but also from light-colored regions of the land mass, such as the polar regions, glaciers, deserts, etc.

The total fraction of incoming radiation reflected back to space, termed the **albedo,** depends mainly on the nature and extent of cloud cover. Until the era of the meteorological satellites it was possible only to make educated guesses about the magnitude of the earth's albedo. At present, however, by directly measuring the radiant energy in the 0.2 to 3 micron region coming up from earth, as sensed by measuring instruments in a satellite orbiting outside the earth's atmosphere,

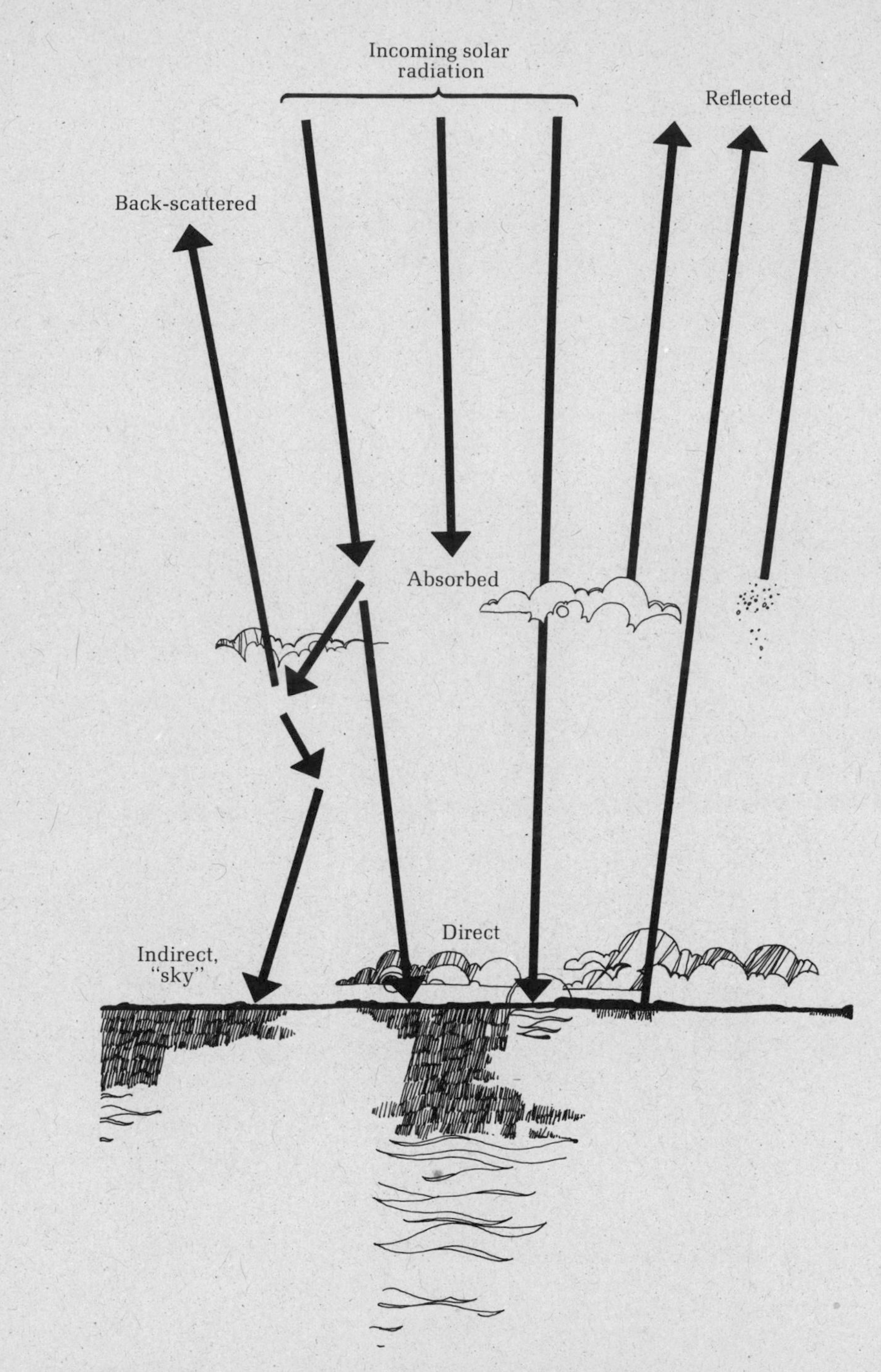

FIGURE 3–1. *Distribution of incoming solar radiation.*

the albedo can be determined directly by comparison with the incoming solar energy. The most recent and extensive measurements of the mean annual global planetary albedo yield a value of 29 percent.[1] This is considerably lower than the earlier estimates based on less direct evidence. Figure 3-2 shows how albedo varies with latitude, as determined from satellite measurements over a two-year period.

The albedo is a very important quantity in consideration of the earth's energy balance because solar radiation returned to space is lost to the planet so far as its energy intake is concerned. It has been estimated that a one percent increase in albedo (that is, from 0.29 to 0.30) would result in a change of about 2°F in the average temperature on the surface of the earth. It seems to be well within man's power to change the albedo, albeit inadvertently, by at least this amount, by altering the amount and type of clouds, the state of the earth's surface, or by changing the amount and type of solid particles in the atmosphere. We will return to this point in a later chapter.

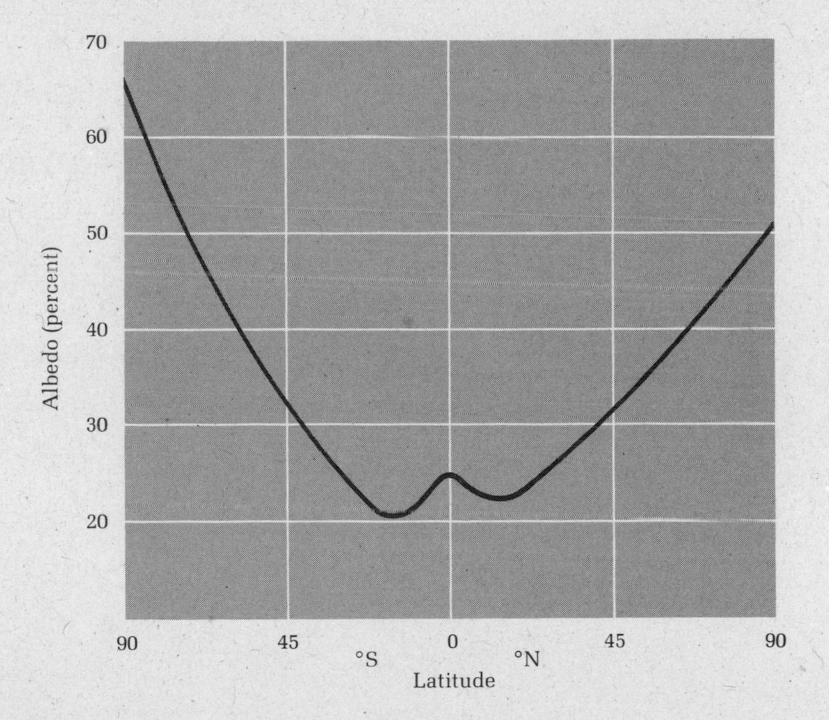

FIGURE 3-2. *Variation in planetary albedo with latitude.*

The atmosphere, in addition to scattering incoming solar radiation, may also absorb it. This absorption, due almost entirely to water vapor, causes some heating. What finally reaches earth, then, is the direct radiation which has gone through unimpeded; and the indirect, sky radiation, which has been scattered one or more times by the atmosphere but finally makes it to the surface. Together these constitute the surface **insolation,** the net amount of solar radiation reaching the surface of the planet.

Now we turn to the other half of the story, the outgoing radiation. Considered in all its details, the terrestrial radiation system is very complex.[2] The most important elements can, however, be seen without looking into all the details. We begin with the earth as a black body radiator at some temperature, say 40°F. The approximation that the earth is a black body radiator for the infrared part of the spectrum is actually quite a good one. As we have already learned, a black body radiating at this low temperature will emit mainly long wavelength infrared radiation. The black body radiation curve for a body at 40°F is shown in Figure 3-3. This radiation leaves the surface and passes upward into the atmosphere. But now we have a situation entirely different from that for the incoming solar

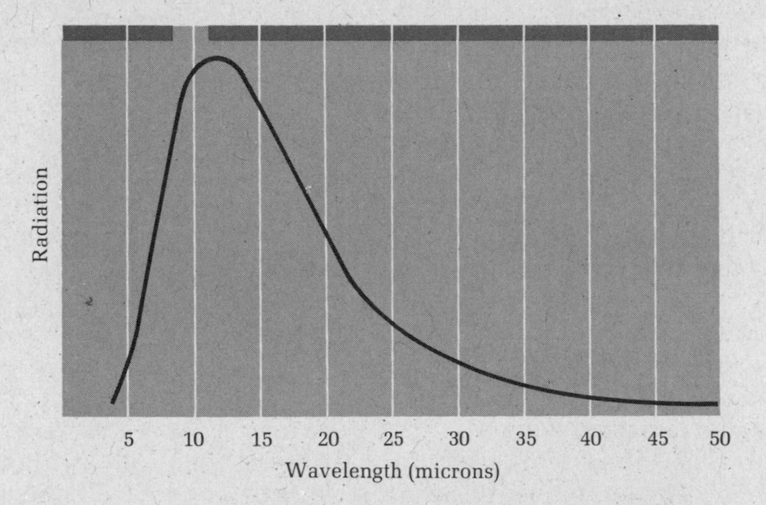

FIGURE 3-3. *The black body radiation of the earth (40°F). The shaded regions across the top represent wavelenghts which are strongly or completely absorbed by the atmosphere.*

radiation. Whereas the atmosphere is mainly transparent for the 0.2 to 3 micron radiation coming from the sun, it is not very transparent at all for the long wavelength radiation leaving the earth. There is a little "window" in the 8.5 to 11 micron range, in which the outgoing radiation passes without strong absorption; everywhere else it is absorbed. The absorption is due to water vapor, carbon dioxide and—to a lesser extent—ozone. The shaded regions along the top of Figure 3-3 mark the wavelength ranges in which atmospheric absorption of the earth's radiation is essentially complete.

We noted earlier that any collection of matter capable of absorbing radiation without reflecting or transmitting it is a black body. In this sense, earth's atmosphere is a black body with respect to much of the long wavelength rays which it receives from earth.

To understand how the absorption of terrestrial radiation occurs, and how it works to establish a temperature distribution, let us consider the atmosphere as divided into layers, as shown in Figure 3-4. It is not too important how thick the layers are imagined to be—say a few hundred feet. Now we must recognize that the atmosphere is not inherently warm. If it were not for the infrared radiation coming up from earth, and the smaller amount of solar radiation absorbed directly, the atmosphere would be extremely cold. It is thus colder on the average than the earth itself. There are, of course, day to day situations in local regions near the surface where the atmosphere is warmer than the surface. Anyone who has swum in cold water on a warm summer day knows this to be true. Nevertheless, in terms of the overall horizontal temperature distribution, and most particularly in terms of the vertical distribution, the atmosphere is clearly colder than earth's surface.

The point of this is that when the first layer of the atmosphere in contact with the surface absorbs the long wavelength radiation from earth, it behaves like a black body absorber. It is also like a black body, however, in having an emissivity. It emits radiation both upward and downward. The earth, therefore, gets some of its radiation back again, but only a fraction of it. The next layer of atmosphere up from the first is now warmed by radiation from the lowest layer, and also by radiation from the surface which passes unabsorbed through the first layer. The total amount of radiation which it receives, however, is less than that experienced by the first layer, and its

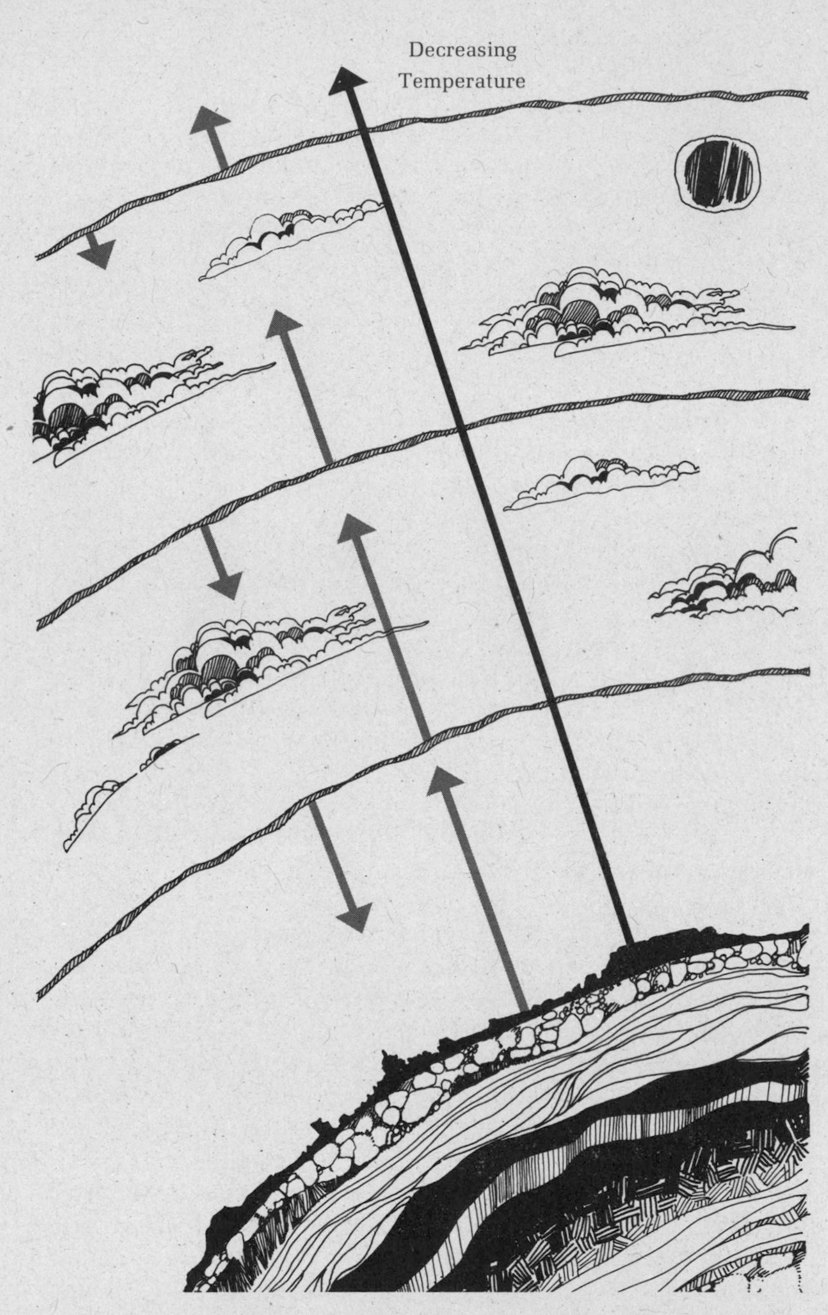

FIGURE 3-4. *Long wavelength (infrared) radiation from earth through the atmosphere. The relative lengths of the gray arrows represent relative amounts of radiation transferred upward and downward.*

temperature is correspondingly lower. It in turn re-radiates, both upward and downward, warming the next layer above it, and returning a fraction of radiation from the lower levels back downward. We can imagine this happening from layer to layer until we arrive out at the edge of a layer in which there is no longer sufficient water vapor or carbon dioxide to absorb the radiation, and it passes off into space.

So far this is fairly simple, once the basic idea is clear. But there are complications. The major infrared absorber in the atmosphere is water vapor. Its concentration varies widely from place to place, and even more widely in vertical distribution. Circulations of the atmosphere may occasionally produce situations in which the usual decrease in water vapor with altitude is reversed, and an upper layer of atmosphere contains more water vapor than a lower. Through a variety of processes, local atmospheric conditions may be formed in which the normal steady decrease in atmospheric temperature with altitude is not observed. There may even be temperature inversions, in which a layer of warmer air lies above a colder lower layer. These occurrences are important in determining weather, but do not change the large-scale picture.

Cloud cover is another very important factor in controlling the vertical temperature distributions. Clouds are composed of liquid water droplets, which are very strong absorbers of infrared radiation. The presence of clouds over a region thus alters the vertical temperature distribution very markedly. The atmosphere under the clouds is warmer than clear atmosphere because of the extensive absorption and reradiation by the clouds of all the infrared radiation reaching them. The atmosphere above the clouds is correspondingly colder. Since clouds reflect radiation from the sun back into space and thus increase the earth's albedo, they function in two opposing ways in determining the temperature of the lower atmosphere and thus of the earth's surface.

The net effect of the clouds depends on their reflectivity and altitude. Detailed calculations by Manabe and Wetherald, based on an elaborate mathematical model for the temperature distribution in the earth's atmosphere, show that low clouds tend to lower the temperature at the earth's surface.[3] They have a high reflectivity, and thus contribute substantially to an increase in albedo. At the same time, because they are low, they have a relatively high temperature, and thus radiate a fair

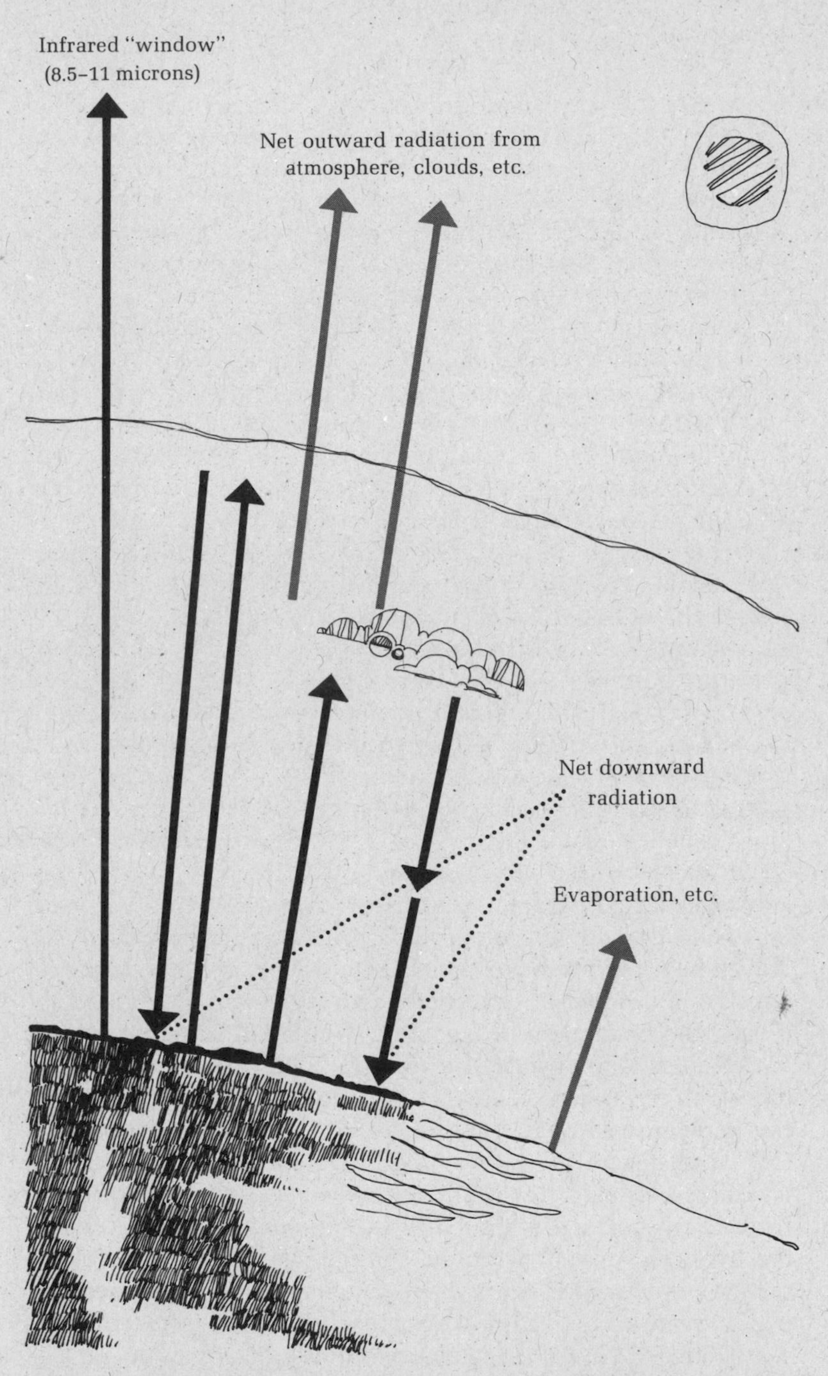

FIGURE 3–5. *Schematic illustration of transfer of heat from earth to the atmosphere and to space.*

amount of infrared radiation into space through their tops. On the other hand, high clouds may tend to increase surface temperature. They have a relatively lower reflectivity toward solar radiation, but are still strong absorbers of infrared radiation. Since they are high, and thus much colder than low clouds, they have a low emission temperature, and thus radiate less infrared energy into the outer atmosphere and beyond.

From even this very sketchy presentation it must be evident that the terrestrial radiation system is complex and involves factors which are not yet fully understood. For our purposes we need recognize only that cloud cover, water vapor, carbon dioxide, and possibly a few other things affect the vertical temperature distribution in the atmosphere, and thus the net long wavelength radiation into space.

The various fates which may befall the outgoing terrestrial radiation are shown schematically in Figure 3-5. In addition to the radiative transfer processes just described, there is also transfer of energy between earth and the atmosphere through the physical hydrologic cycle, i.e., the movement of water via precipitation, condensation and evaporation. Figure 3-5 depicts processes which are, in a sense, complementary to those shown in Figure 3-1. The earth's surface reaches some temperature such that the net outgoing long wavelength radiation equals in total energy content the net incoming solar radiation absorbed by the surface.

Measuring instruments mounted in satellites and directed toward earth have been employed to measure the long wavelength radiation from the planet. By employing measurements at a number of different wavelengths in the infrared region, it has been possible to obtain detailed information on the variation of atmospheric temperature with altitude.[4] One such temperature profile is shown in Figure 3-6. The overall, effective temperature of the planet, as seen from outside, is quite low, on the order of -50 to -100°F. The amount of outgoing infrared radiation varies with latitude, and is somewhat higher from the tropical belt than from the poles. Figure 3-7 shows the averaged annual infrared emission as a function of latitude, as determined from satellite measurements.[1a] Averaged over a year, and over the various latitudes, the infrared emission from earth is 0.33 langley per minute.

We might now ask whether the values reported from satellite measurements for the solar constant, the earth's albedo and the infrared emission from earth are all consistent. The radiation balance for the earth requires that, if the earth is not to be steadily heating up or cooling down, the incoming radiation which gets through equals the infrared radiation coming out, averaged over a reasonable period of time, a year or so. If we are to compare the measured quantities, however, we must first make a calculation. Recall that the solar constant is the radiant energy from the sun which falls per minute on a unit area of one square centimeter, perpendicular to the sun's rays.

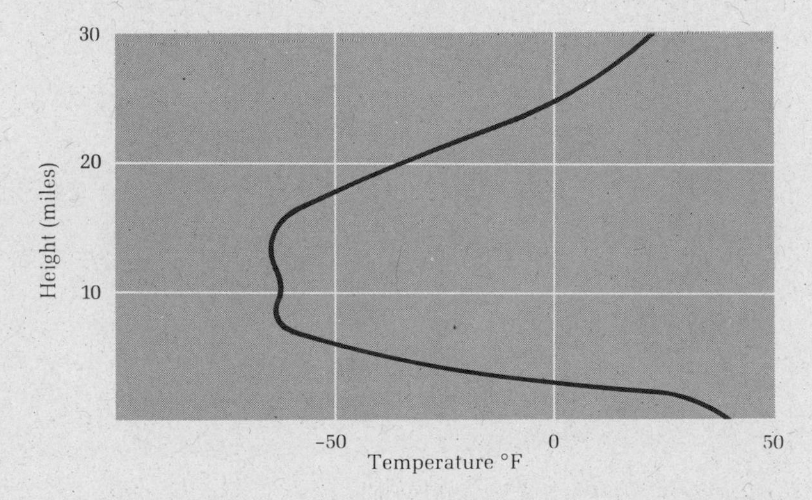

FIGURE 3-6. *Atmospheric temperature profile above Green Bay, Wisconsin, April 18, 1969, with clear skies, at approximately 11 P.M. The data for development of the temperature profile were obtained from a Nimbus III satellite.*

But no part of the earth is perpendicular to the sun's rays for more than an instant in any day, and some parts never are. If we calculate the *average* solar radiation falling on a unit area of the earth, averaged over all of the earth, for a full day, including the hours of darkness, we obtain 0.48 langley per minute. To correct this for the albedo, we multiply by (1 – albedo), since this fraction represents the part which is absorbed at the earth's surface or in the atmosphere. The albedo is 0.29; we therefore obtain 0.34 langley per minute for the averaged value of the insolation, the solar radiation entering each unit area of the earth's surface. This is to be compared with the measured value of 0.33 langley per minute for the infrared radiation outward. The slight difference is certainly within the experimental uncertainty of the measurements. Thus, these very elegant satellite measurements have established that earth is indeed in radiation balance, or very nearly so.

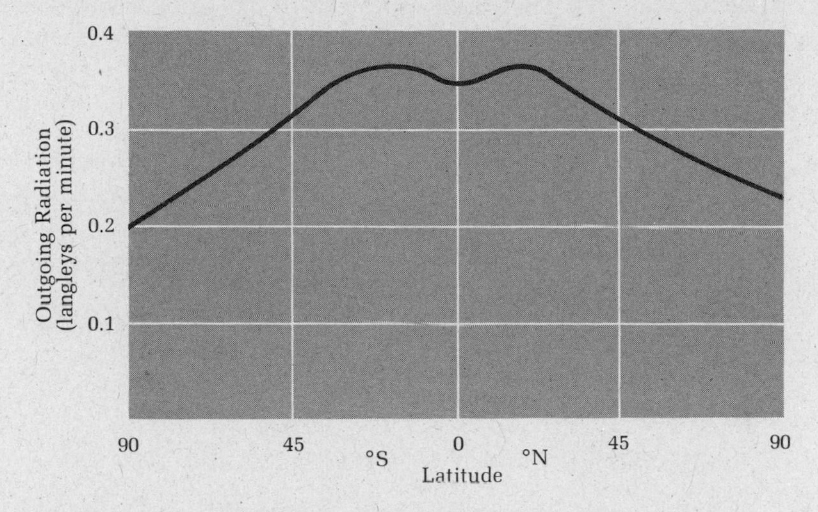

FIGURE 3-7. *Variation in outgoing planetary radiation with latitude.*

But if the earth's radiation budget as a whole is balanced, there are very sizable inequities based on latitudinal distribution. We can see this by comparing Figures 3-2 and 3-7. Figure 3-7 shows that infrared emission from earth varies somewhat with latitude, but the variation is not very marked. Thus, the polar regions emit radiation into space at a rate more than half as great as in the tropics. On the other hand, the incoming solar energy is very heavily concentrated in the tropical belt, between 30° north and 30° south. This is so because this region of the globe presents the largest cross-sectional area of the sun's rays (refer to Figure 2-1), and because the albedo of the polar regions is so high, as shown in Figure 3-2. Very little of the solar radiation which does fall on the polar regions is absorbed.

But if essentially all of the incoming solar radiation is absorbed in the tropical zone, and radiation to space is more uniformly distributed over the earth's surface, there must be a world-wide energy distribution system in operation. Heat transfer must occur from equatorial regions to the higher latitudes.

The ocean and atmosphere together constitute the machinery by which this transfer of energy is effected. Since most of the surface in the tropical regions in which the solar energy is absorbed is water, the ocean-atmosphere interaction is very important. Most of the energy moves from the ocean to the air by evaporation of water from the ocean surface. A certain amount of heat is required to evaporate a liquid, i.e., to convert it into a gas. When water evaporates from the ocean surface, it carries with it into the atmosphere the energy required to cause the evaporation. It has been estimated that about one trillion (1,000,000,000,000) tons of water are evaporated, condensed into clouds and released as precipitation in the earth's atmosphere during a typical 24-hour period.

The energy brought into the tropical atmosphere through water evaporation is transferred to upper layers of the troposphere where it is transported to higher latitudes by fast-moving air currents. The large-scale movements of energy and water vapor, and the continuous interaction of the atmosphere and oceans provide the driving force for the great oceanic currents and trade winds, and determine the broad outlines of the earth's climate.

4 the invisible blanket

We have seen how atmospheric temperature is related to the absorption of long wavelength planetary radiation. It has often been remarked that the atmosphere is much like the glass walls of a greenhouse, which are transparent to the incoming short wavelength solar radiation, but retain (through absorption and reradiation) the long wavelength infrared rays. The atmospheric walls of our terrestrial greenhouse are more impalpable, but nonetheless similarly effective in keeping the earth warm.

The principal infrared absorbers among the components of the atmosphere are ozone, water vapor and carbon dioxide. Figure 4-1 shows the regions of the infrared spectrum in which each of these gases absorb strongly. Ozone does not play a large role in the atmospheric greenhouse effect because its infrared absorptions are not very extensive, and its total presence in the atmosphere is not large. Water vapor is the major component, both in terms of the strength with which it absorbs, and in terms of its total presence in the atmosphere. By comparing the absorption regions with the black body radiations from the earth, however, we can readily see from Figure 4-1 that carbon dioxide might be important in atmospheric absorption of the long wavelength radiation.

Carbon dioxide is one of the minor constituents of the atmosphere (Table 1, page 7). It is a colorless, odorless, nontoxic gas, most commonly seen as bubbles in carbonated soft drinks, beer and champagne. Carbon dioxide molecules consist of one atom of carbon and two of oxygen: CO_2. The substance is present in large total quantities in the oceans of the world. A great deal more of it is chemically trapped, so to speak, in minerals in the form of carbonates. Combustion of any carbon-containing substance such as coke, charcoal, coal, gas, oil or wood in the presence of adequate amounts of air leads to formation of CO_2. It is formed also in animal metabolism, which is really just a form of controlled combustion. We inhale air, and our lungs extract some of the oxygen from it. We exhale air which has been enriched in carbon dioxide. Plants, on the other hand, take in carbon dioxide. Through photosynthesis they convert carbon dioxide, water and other nutrients into oxygen and plant matter.

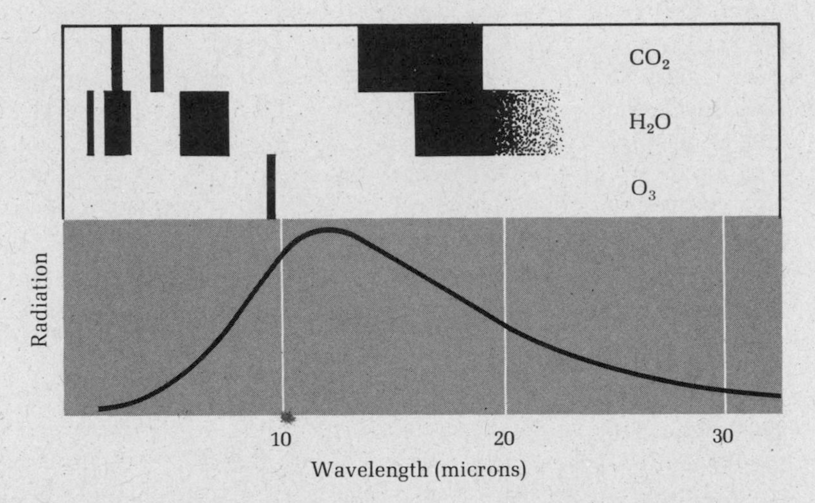

FIGURE 4-1. *Infrared absorptions of carbon dioxide, water and ozone, as compared with the long wavelength radiation from earth.*

The possible importance of carbon dioxide in atmospheric absorption was first recognized more than a hundred years ago. John Tyndall, a British physicist, pointed out that if there were substantial changes in the quantities of infrared-absorbing gases present in the atmosphere, drastic changes in climate might result. After Tyndall's work, a number of scientists performed various calculations to show that variation in the carbon dioxide level in the atmosphere could be responsible for the climatic variations of the past. These attempts to account for the climatic history of the earth have met with competition from a number of other theories, based upon supposed variation in some single other factor which might be responsible for varying temperature conditions on the planet, for example, volcanic dust or perturbations of the earth's orbit.

More recently, a new concern has arisen. Through the combustion of carboniferrous fossil fuels (coal, oil, natural gas), man is adding significantly to the carbon dioxide content of the atmosphere. It is a matter of more than passing interest to know, therefore, to what extent variation in the carbon dioxide level in the atmosphere may be expected to change the atmospheric and surface temperatures of the planet.

It sounds quite plausible that if carbon dioxide is an infrared absorber, increasing the content of carbon dioxide in the atmosphere should result in an increase in the "greenhouse effect" and the earth's temperature should go up. Assuming that this intuitive conclusion is correct, the question then becomes: how much should it go up for a given increase in carbon dioxide level? To answer this quantitative question is no easy matter; yet, it is obviously important that we have an answer. There are no experimental paths open to us; we cannot suddenly double the carbon dioxide of the atmosphere and then observe what happens over a period of time. In the long run, as we will see later, man might carry out the "experiment" quite inadvertently, and with possibly disastrous consequences. What we need for the present is some basis on which to judge what the effects of increasing the carbon dioxide content *would* be. If it can be reliably shown that the effects would be exceedingly unpleasant for the inhabitants of the planet, the long-range "experiment" might be called off in time.

We must, then, have recourse to theoretical models in predicting the effects of changes in composition of the atmosphere, and in particular, changes in the carbon dioxide level.

To understand the CO_2 problem properly we must first recognize that carbon dioxide and water vapor compete with one another to some degree as absorbers of the earth's infrared radiation. Carbon dioxide has its major absorptive power in the range from 12 to 18 microns (Figure 4–1). A major region of water vapor absorption extends from about 15 microns to longer wavelength. In the spectral range from about 15 microns outward, therefore, both water vapor and carbon dioxide may absorb. A second point of importance is that these two infrared absorbers are not similarly distributed throughout the atmosphere. Carbon dioxide is quite uniformly present, to the extent of about 318 parts per million, at the surface and at all altitudes above it. Thus, as the atmosphere grows thinner at higher altitudes, the carbon dioxide concentration diminishes along with it.

By contrast, water vapor is not at all uniformly distributed. The water vapor pressure is highest at the surface, and diminishes very rapidly with increasing height. Although the distribution varies much from place to place, and with time, it is invariably the case that thc water vapor pressure is only on the order of a few parts per million at the tropopause. Figure 4-2 shows graphically how the two gases compare in terms of this vertical distribution. At the surface, assuming normal relative humidity conditions of 75 percent, and an air temperature of 80°F, the water vapor content is on the order of 30,000 parts per million, on a volume basis. This is enormously larger than the carbon dioxide level. At an altitude of a few miles and above, however, carbon dioxide is more abundant than water vapor. Any serious attempt to calculate the effect of carbon dioxide on the atmospheric temperature must properly account for this different distribution of carbon dioxide and water vapor. Cloud cover is another factor which requires careful consideration.

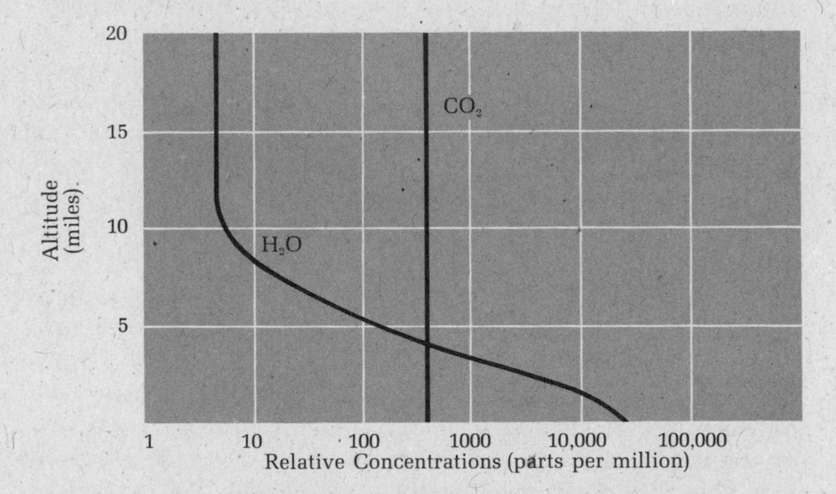

FIGURE 4-2. *Variation in water vapor and carbon dioxide concentrations with altitude.*

Since the 1950s, beginning with the work of G. N. Plass,[1] a number of attempts have been made to calculate the effects of changing carbon dioxide level on the atmospheric and surface temperatures, when equilibrium conditions prevail. These attempts have all been subject to criticism on various grounds. To solve the problem with the computational tools available at the time, it was necessary to adopt a model for the earth-atmosphere system which contained too many simplifications of the real situation. The most recent work, by S. Manabe and R. T. Wetherald, is based on the most realistic mathematical model for the atmosphere yet developed for work of this kind.[2] It takes proper account of the vertical distribution of water vapor, carbon dioxide and ozone. It includes such features as surface albedo and variable cloudiness. It allows for the attainment of an equilibrium temperature distribution in the atmosphere by the radiative processes described in Chapter 3 and illustrated in Figure 3–4. In addition, it allows for radiative convection, which prevents the formation of excessively large temperature changes with change in altitude.

The procedures used in carrying out such a calculation are something like this: first, the equilibrium distribution of atmospheric temperature is calculated for an atmosphere which contains 300 parts per million of carbon dioxide, and in which the relative humidity[3] is assumed to vary with altitude. The assumed variation in water vapor content with altitude is chosen to match closely the available experimental evidence. Thus, if a given relative humidity at the ground is chosen, the water vapor content in the atmosphere depends on the temperature profile finally arrived at in the calculation. An amount of cloudiness which represents the average for the planet is assumed. The calculations are then carried out on a large computer. The final outcome is a temperature profile for the atmosphere. It is very encouraging that the temperature variation with change in altitude predicted by the calculation is very much like the typical profile observed, as depicted in Figure 3–6. In the second stage the calculation is repeated using a different carbon dioxide level, say 600 parts per million. We may assume that this will yield a different temperature distribution in the atmosphere. But now there arises a difficult problem: should we assume that the amount and distribution of water vapor is the same as in the previous calculation, or should we assume that the amount of moisture varies with the air temperature? Assuming constant *relative* humidity is equivalent to making the latter assumption, since, for a constant relative humidity, warmer air contains more water vapor, and colder air less. The more reasonable expectation is that the relative humidity will remain constant as atmospheric temperature varies.

When the calculations are carried out with this assumption, and for an assumed 600 parts per million level of carbon dioxide, a new equilibrium atmospheric temperature profile is obtained. This profile has higher temperatures in the troposphere and lower temperatures in the stratosphere than the previous result. Thus, the climate, as determined by the surface and troposphere temperatures, is made warmer by the increase in carbon dioxide level. Manabe and Wetherald calculate an increase of 4.2°F for the equilibrium surface temperature for a doubling of carbon dioxide level, assuming constant relative humidity for the atmosphere.

We can understand what happens in this way: Much of the earth's infrared radiation is absorbed by the lower levels of the atmosphere, and reradiated both upward and downward, as described in the previous chapter. Cloud cover affects the temperature distribution also, by its very strong infrared absorption. A little of the absorption and reradiation in the lower atmosphere is due to carbon dioxide, but water vapor plays the predominant role because its concentrations are so high.

In the higher atmosphere, where water vapor levels are very low, radiation upward is intercepted mainly by carbon dioxide. There is enough of this radiation, particularly over the regions of the earth free of cloud cover, to make the absorption and reradiation by carbon dioxide a significant factor. Now if the carbon dioxide level goes up, the reradiation downward from the higher altitudes increases, warming the troposphere. But if the troposphere warms, and if we assume that there is a constant *relative* humidity, a kind of amplification effect sets in. The warming puts more water vapor into the atmosphere, which in turn means a stronger absorption of infrared radiation in the lower atmosphere. The result, according to the calculations, is that the temperature increase resulting from doubling of carbon dioxide level is twice as great, assuming a constant relative humidity, as it is assuming that there is no change in the water vapor level with increasing atmospheric temperature. As we have already noted, the assumption of a constant relative humidity seems in best accord with what is presently known about the atmosphere.

The calculation of Manabe and Wetherald is by no means the last word on the subject. Although it takes into account many aspects of the real atmosphere which were neglected in previous work, the model which they employ still omits many characteristics of the planetary atmosphere which might be important in determining the magnitude of the carbon dioxide effect. We can probably assume with some safety, however, that the results of more sophisticated future calculations,[4] requiring larger computers than were available before, will not drastically change the value they have obtained.

But if we have an estimate for how much the average temperature of the earth's surface would change for a doubling of carbon dioxide level, the next question is: how much can we expect the carbon dioxide levels to change in the future as a result of man's activities? And when we have an estimate of this and can estimate the total temperature change which might occur, what will we be able to make of this? What does a given change in the average temperature of the earth's surface mean in terms of climatic conditions?

There are about 2,500,000,000,000 tons of carbon dioxide in the atmosphere (huge numbers of this sort are best written in exponential form: 2.5×10^{12} tons). A much larger quantity of carbon dioxide, 140×10^{12} tons, is dissolved in the oceans. The total amount of CO_2 tied up in the biomass, i.e., the amount which would be released if all plant and animal life were completely decomposed, is comparatively small, about 0.5×10^{12} tons. It is with these quantities in mind that we must estimate the effect which man's burning of fossil fuels might have on the carbon dioxide level in the atmosphere. It is difficult to obtain reliable estimates of the total recoverable world fuel reserves. Much of the world has still not been thoroughly assayed for its mineral and fuel content. Secondly, new technology may make possible the recovery of fuels from sources which do not now appear very promising. In spite of these uncertainties, however, estimates of world-wide reserves, recoverable in terms of present technology, have been made.

The major source of fuel is coal. The total recoverable coal reserves are estimated to be about 4.7×10^{12} tons, on the basis of data recently summarized by Hubbert.[5] The crude oil and natural gas reserves can be expected to yield a total of about

0.35×10^{12} tons coal equivalent. Thus the total expected recoverable reserves can be set at about 5×10^{12} tons of coal equivalent. This estimate, which does not include possible contributions from oil shale, tar sands, and other low grade sources, is lower than some others, which range to more than twice this. At the Seventh World Energy Conference, held in Moscow in August, 1968, overall geological reserves of fossil fuels were estimated at from 10×10^{12} to 25×10^{12} tons,[6] of which 3.4×10^{12} tons can be extracted at an economically justified cost. Presumably the latter estimate is made in terms of present economic considerations. As fuel becomes scarcer, some of the less desirable sources will be worked. Our estimate of 5×10^{12} tons therefore seems conservative.

The major fraction of oil and gas contributions to this total will have been exhausted in another 60–70 years. Coal production may be expected to reach a maximum by about 2100, with consumption at this time of about half the total reserve. About 90 percent of the 4.7×10^{12} tons of estimated recoverable coal will, according to the predictions, have been consumed by 2350–2400.

Combustion of a ton of fossil fuel, whether of coal, oil or gas, produces about 3.2 tons of carbon dioxide. The combustion of all the estimated recoverable fossil fuel will therefore produce about 16×10^{12} tons of carbon dioxide. This is 6.5 times the total amount now present in the atmosphere. If it were all to remain in the atmosphere, and if the Manabe-Wetherald estimate of temperature increase were to hold for such a large increase in carbon dioxide level, the average temperature increase would be 27°F! Fortunately, a change of this magnitude, which would be catastrophic in terms of climatic change, seems out of the question, as we shall now show.

Carbon dioxide has one important characteristic which distinguishes it from most other substances inadvertently added by man, or sporadically added by nature (for example, volcanic dust). Because it is not very active chemically, it does not readily undergo change to some other substance which might more rapidly drop out of the atmosphere. Furthermore, it is quite out of the question that man could devise some method for removal of carbon dioxide from the atmosphere if this should turn out to be desirable. The quantities involved and the vast extent of the entire atmosphere rule out this prospect.

Studies have shown that plant growth increases in the presence of increased carbon dioxide concentration. We should consider, therefore, whether the increased amount of carbon dioxide in the atmosphere might not simply be taken up in the form of increased plant growth over the planet. One factor which works against this is the clearing of land by man, thus eliminating significant areas of heavy forest growth, etc. On the other hand, there might be a greatly increased growth of phytoplankton in the oceans, perhaps stimulated by the increased flow of man-made nutrients into the oceans. But even if there were an increase in the total biomass, this could not result in a significant removal of carbon dioxide. Plants which remove carbon dioxide from the atmosphere eventually die and decay, releasing carbon dioxide back into the atmosphere-ocean system. Since the amount in the total biomass is only about 20% of the present atmospheric level of carbon dioxide, a large percentage increase in the total biomass, say 50%, would remove only about 10% of the present atmospheric carbon dioxide, and an even smaller percentage of the increased levels expected in the future.

Much of the carbon dioxide added to the atmosphere will be absorbed by the oceans (Figure 4-3). The first stage in the exchange of atmospheric and dissolved carbon dioxide involves the air-sea interface. In recently reported experiments,[7] R. Berger and W. F. Libby found that when a certain quantity of carbon dioxide gas was placed in contact with a large quantity of turbulent surface seawater, half of it dissolved in the water in a period of about 470 days. Thus, the exchange of carbon dioxide with surface seawater is not extremely rapid. Carbon dioxide is present in seawater in various forms. Because it is a weakly acid substance in water, forming carbonic acid, the amount of CO_2 which can be held in slightly acid water is less than for more alkaline water. Since carbon dioxide makes water more acidic simply by dissolving, it acts to some degree as an inhibitor of its own solubility. It is possible to carry out calculations which would give some idea of how much of the carbon dioxide in the atmosphere will finally end up in the oceans. In the long run, i.e., over several thousand years, most of it will. The difficulty with such calculations is that they assume the ocean-atmosphere system to be in equilibrium. There is very good reason for believing that such a complete equilibrium is not attained for thousands of years. It is well established that the deep waters of the ocean exchange with the sur-

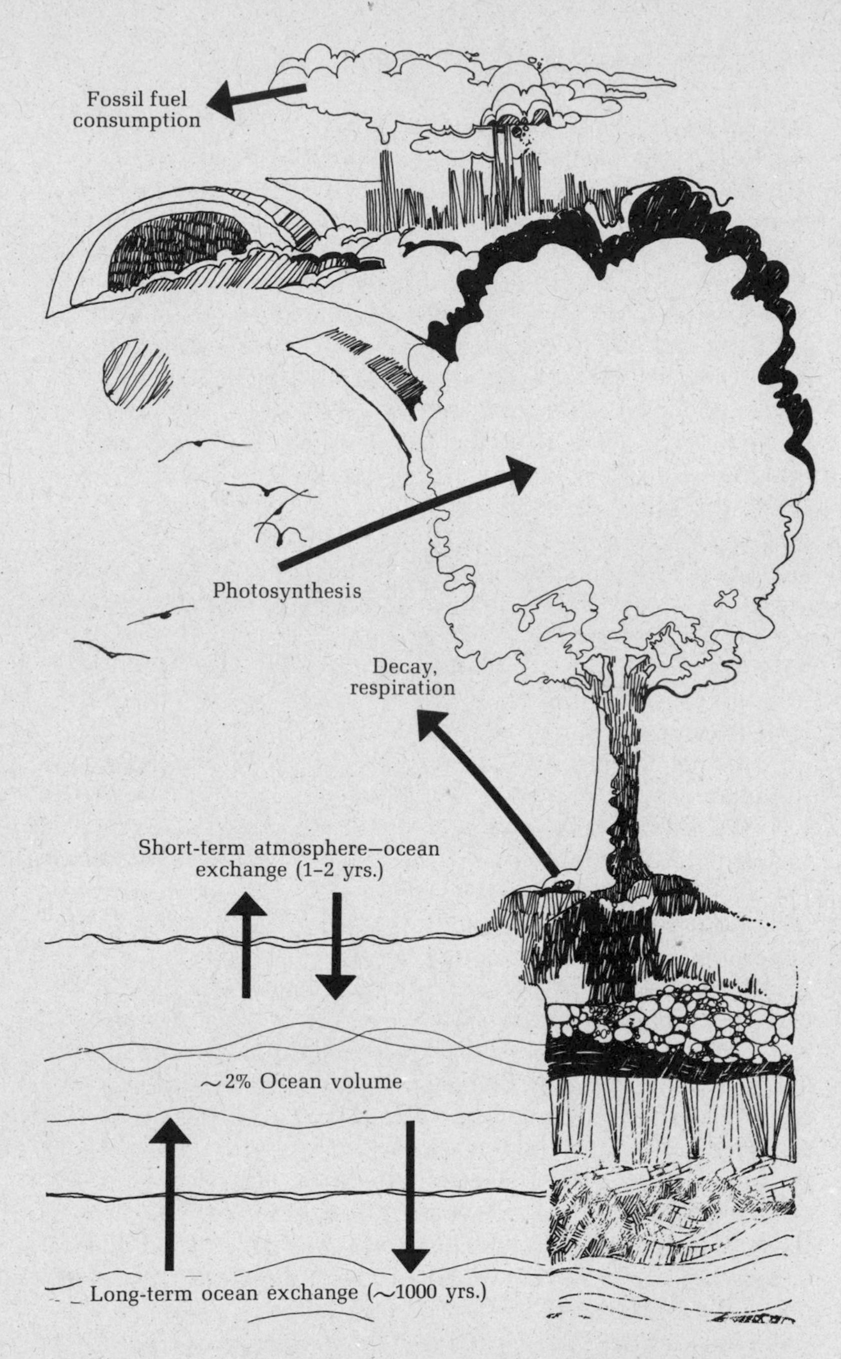

FIGURE 4–3. *Simplified diagram of carbon dioxide distribution and exchanges. The tree represents all plant and animal consumption and release of* CO_2.

face waters only at a very slow rate.[4] Radiocarbon data indicate that the average deep water molecule in the ocean makes its way to the surface once each thousand years or so.[8] In considering the exchange of atmospheric carbon dioxide with the oceans, therefore, only the upper layers of the oceans, containing about 2 percent of the total volume,[9] are important on a time scale which is related to our concerns.

Although there is some variation in the quantity of dissolved carbon dioxide in a unit volume of ocean water as a function of depth,[10] we can assume without too much error that this 2 percent of the ocean which exchanges carbon dioxide with the atmosphere contains about 2 percent of the total dissolved carbon dioxide, i.e., $0.02 \times 140 \times 10^{12}$ tons $= 2.8 \times 10^{12}$ tons. This is just a little more than the amount of CO_2 present in the atmosphere. As a rough guess, therefore, we could say that if the carbon dioxide content of the atmosphere were suddenly to double, somewhat more than half of it would dissolve in the oceans within a few years, and most of the remaining extra would dissolve extremely slowly, over a period of several thousand years.[11]

Man is not, of course, adding the carbon dioxide to the atmosphere suddenly, but rather at a steady, slowly increasing rate. We can guess on the basis of the picture we have just developed that about half of the added CO_2 is being dissolved in the oceans at nearly the same rate as it is being produced, and that the rest will remain in the atmosphere as increasing carbon dioxide level for a long time to come.

Fortunately, these rough estimates can be tested with real experimental data. C. D. Keeling and his co-workers, associated with the International Meteorological Institute, Stockholm, Sweden, and the Scripps Institution of Oceanograpohy made accurate measurements over a few-year period of the carbon dioxide level in the atmosphere at selected locations on the globe.[12] The most extensive data were accumulated at the Mauna Loa Observatory, Hawaii. Figure 4-1 shows the results for the period 1959–1962. These and similar data[11,12] show a steady increase in carbon dioxide level amounting to 0.7 parts per million per year.

Thus, during this period, the net amount of carbon dioxide added to the atmosphere in a year is given by the amount of increase in one year, 0.7, divided by the total amount present in the atmosphere, say 314, times the total weight of all that present, 2.5×10^{12} tons. The result is 5.5×10^{9} tons of carbon dioxide. Now we need an estimate of how much carbon dioxide was added to the atmosphere during this time.

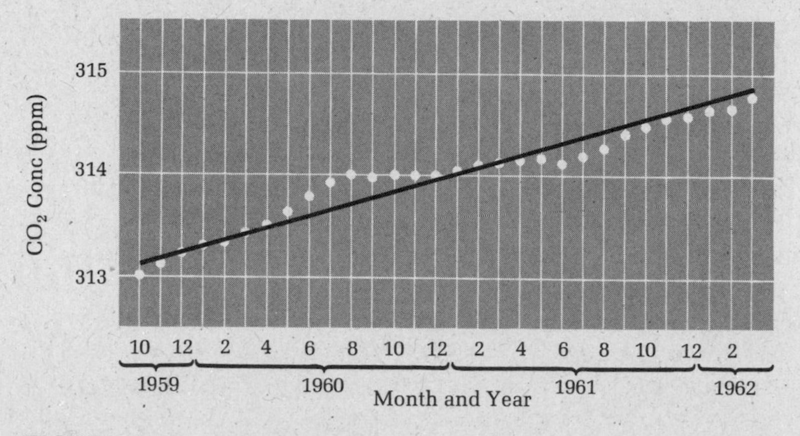

FIGURE 4–4. *Long-term increase of carbon dioxide in the atmosphere. Average monthly concentrations measured at the weather station on Mauna Loa in Hawaii.*

At the Seventh World Energy Conference held in Moscow it was reported that the world consumption of energy has doubled in the past sixteen years.[6] A recent United Nations report places the total world consumption of energy for 1966 as 5.51×10^{9} tons of coal equivalent. If we make allowances for the growth between 1960 and 1966, and take off about 4% for energy derived from sources other than fossil fuels, we obtain an estimated consumption of about 4.6×10^{9} tons of coal equivalent for 1960. This converts to about 15×10^{9} tons of carbon dioxide, to be compared with 5.5×10^{9} tons observed for the atmospheric increase. Thus, it appears from these data that about 37% of the carbon dioxide being put into the atmosphere is remaining there to add to the long-term increase in the level of the gas. This figure is not too far from our rough guess that somewhat less than half would remain in the atmosphere. Returning to the original figure of how much carbon dioxide would be released by combustion of *all* the fossil fuel reserves, we must now reduce this by the factor 0.37 to obtain a more realistic estimate of the change in carbon dioxide level, and the consequent change in average temperature. We anticipate that the carbon dioxide level would be on the order of 3.4 times larger than at present, and that the change in average temperature would be on the order of 10°F.

Putting aside for a moment questions about whether man may indeed be expected to use up the quantities of fuel which would produce this large a temperature effect, let us ask what the consequences of a change of this magnitude in surface temperature might be. There would necessarily be an increase in the pole-ward flux of heat from the tropical regions. Since the added carbon dioxide increases the surface temperature by inhibiting the outward flow of long wavelength radiation, and since this radiation is more intense in the tropical than in the polar regions, the influence of carbon dioxide will be felt most greatly in the tropical regions. This means then that a substantial transfer of the added heat to the polar regions should result. It is not possible to do more than conjecture about whether the changes in air mass movements would result in substantial changes in climate for large portions of the presently habitable land areas. Meteorologists and climatologists believe that within the past few thousand years there have been significant changes in temperature for periods as long as 1,000 years or so. These changes, whose origins are not clear, have resulted in very drastic changes in the extent of rainfall, and thus in the nature and extent of plant and animal life which a region can sustain. For example, in the period around 4000 BC there was a change to a warm, humid climate throughout Europe and North America. Average temperatures may have been as much as 5°F warmer than at present, with heavy rainfall, and growth of peat. Judging from flora deposited in the arctic during that period, the northern polar region was free of ice, i.e., there was no north polar ice cap. Assuming that this climate was widespread, there was probably very little glacial coverage anywhere on the planet.[13]

Since this period, termed the Climatic Optimum, there have been many pronounced changes in the climate over Europe, including periods of dry heat lasting for hundreds of years, and changes to a colder, wetter climate. All of these climatic alterations arose from one or more causes unknown to us. The point is that changes in climate which are very pronounced in terms of rainfall distribution and other weather factors may accompany, and may in fact result from, temperature variations of only a few degrees.[14]

It has been remarked in the report of a National Academy of Sciences-National Research Council panel[15] that the climatic changes expected to result from increased carbon dioxide levels are no greater than those experienced by Europe and North America since the last glaciation, and "although some of the natural climatic changes have had locally catastrophic effects, they did not stop the steady evolution of civilization."

That the evolution of civilization has been steady is certainly a debatable contention. More importantly, however, man lives on a planet which is rapidly becoming too small. For an increasing world population, facing the prospect of ever increasing food shortage, a major climatic change which would materially reduce the availability of land for raising food and sustaining civilized life would be a disaster of unparalleled proportions. In this connection, one possible consequence of the increase in average temperature is a steady melting of glacial fields and the polar ice. Melting of glaciers would have the effect of decreasing the earth's albedo, since the ice has a higher reflectivity than the land beneath. This in turn would result in some further increase in the earth's temperature, thus constituting an amplification of the carbon dioxide effect. (Another self-amplifying effect arises from the fact that the solubility of carbon dioxide in the oceans decreases with increasing temperature. An increase of 4°F in the ocean temperature would result in a 12 percent increase in carbon dioxide of the atmosphere due to CO_2 expelled from the water.[16]) Secondly, melting of glaciers and of the antarctic ice would result in an increase in the volume of the oceans. It is estimated that melting of all the ice would cause an increase of about 200 feet in sea level. Obviously, such a rise would put much of what is now heavily populated land under water.

Melting of the Antarctic ice would occur only slowly, during a period of a few thousand years, and probably would not proceed to a major extent before much of the added carbon dioxide had dissolved in the oceans. Melting of the glaciers and of the Arctic ice cap would, however, be much more rapid. The Artic ice is afloat, so that its melting would not result in an increase in the water level of the oceans. It is reasonable to expect, however, that in the absence of polar ice there would be major changes in the oceanic currents in the northern hemisphere, with consequent changes in climate of the land areas. The new climatic regime, while bringing a more temperate climate to the subpolar regions, could result in considerably drier weather for many other parts of the world.[14] We do not as yet have a sufficient understanding of the planet's "climate machine" to be able to predict with any reliability what the good and bad effects of a major alteration in the nature of the earth's surface, such as melting of the polar ice cap, might be. In the absence of such a predictive ability, it seems only prudent to avoid forcing any such major alterations.

Those who are skeptical of the prospective importance of the changing carbon dioxide level in the atmosphere point out that since about 1940 the average temperature of the planet has apparently decreased slightly, about 0.4°F, whereas during this time the carbon dioxide content of the atmosphere has been increasing. Since the mid-1930s man's activities have resulted in the addition of approximately 0.30×10^{12} tons of carbon dioxide to the atmosphere, a very generous estimate. Applying the correction factor for dissolution in the sea, we obtain 0.11×10^{12} tons of added carbon dioxide in the atmosphere. This is about 4.5% of the total atmospheric carbon dioxide. Using Manabe and Wetherald's estimate, this should therefore have produced a temperature change of only about 0.18°F during this time. Obviously, this is a small amount, and it is not difficult to imagine that other factors may affect the average temperature to this degree. Man's use of fossil fuels is, however, increasing at a very rapid rate, and with the passage of time, ever larger amounts of carbon dioxide will be deposited in the atmosphere. If, for example, the use of fossil fuels continues to increase at the present rate of increase for energy as a whole, i.e., with a doubling time of 16 years, by the middle of the next century man will have consumed about 80% of all the estimated recoverable reserves! While no one anticipates that the use of fossil fuels will be this intensive, it is quite likely that 80 percent of all the estimated crude oil and natural gas reserves, and about 30 percent of the estimated recoverable coal reserves will have been converted to carbon dioxide. This will result in just about a doubling in the atmospheric carbon dioxide, after allowance for the short-term equilibration with the oceans. We may thus anticipate that added carbon dioxide will result in an increase of about 4°F in the earth's average temperature by about the year 2050. Furthermore, unless there are radical changes in the approach to energy conservation, it seems quite probable that continuing generation of carbon dioxide will be very substantial at that time. The implications seem clear enough; carbon dioxide will be an important factor in changing the nature of the earth's climate during the next century or two. Future generations may suffer greatly from those changes.

5 red at night, sailors delight

The atmosphere consists almost entirely of the gaseous components listed in Table I, page 7. The remainder is solid particulate matter, present in very small quantities in the troposphere and to an even lesser degree in the stratosphere. The total mass of all the solids in the atmosphere is probably about 20 million tons or more. This seems a rather substantial amount, but it is only a few parts per billion of the total atmosphere. A component present in such extremely low concentrations would not be worth discussion except that it plays a vital role in shaping weather and climate, on both a regional and a world-wide basis.

The term **aerosol,** which is most commonly applied to a dispersion of tiny liquid droplets in the air, applies also to a semi-permanent suspension of tiny solid particles. If the particles are to remain suspended in the air for any appreciable time, they must be small. Although there may be some variation occasioned by differences in density, smaller particles are usually carried to higher altitudes and remain in the atmosphere longer than large particles. Those which remain suspended in the air for long enough to matter in our discussion must be no larger than about one micron in diameter. (Recall that a micron is a millionth of a meter, or about 0.000038 inches.) Particles important in affecting weather may be as small as only 1/100 micron in diameter.

The atmosphere receives aerosol particles from smoke of all kinds. Some of the sources, such as volcanic eruptions and forest fires started by lightning, are entirely natural in origin. Others, such as smoke from combustion processes and forest fires started by careless campers, are man-made. Much of the solid debris generated by all these processes consists of rather large particles which remain at relatively low altitudes for a short while before being carried to earth by rain or a snowfall. The submicron particles, however, may be carried aloft by convective air movement. Volcanic eruptions may generate updrafts extending into the stratosphere which carry the smaller particles aloft.

Ocean water spray, formed by wave action, consists in part of tiny water droplets which are small enough so that they are carried in the atmosphere for considerable time. During that time the water may evaporate, leaving microscopic salt crystals suspended in the air. Near the ocean the atmosphere in the first few thousand feet contains in each cubic foot several million salt particles of about 0.1 micron diameter. These smaller components of the salt aerosol may remain in the atmosphere for many weeks, and be carried far from the place of origin.

Solid particles may also be blown into the air from the surface of the earth. Clearing the land for agricultural purposes and stripping forests in lumbering activities have undoubtedly resulted in increased atmospheric dustiness. Much of the dust may remain in the atmosphere for days and be carried over many miles. The dust storms which occurred throughout the Oklahoma-Kansas area during the 1920s and 1930s, for example, resulted in heavy fallouts of dust in rainfall over the eastern U.S.

An additional, although minor, source of atmospheric particles is the influx of meteoroids from outer space. Each day hundreds of millions of meteoroids enter the earth's atmosphere. Most of these are vaporized by the heat generated during their passage at high velocities through the atmosphere. Those of submicron size slow very quickly and drift down to the surface.

The cycle of water from oceans into atmosphere into cloud formation and subsequently into precipitation is one of the most important aspects of the earth's climate. Solid aerosol particles play an essential role in cloud formation, and in triggering precipitation. The formation of rain or snow from water vapor requires three distinct processes. First, the air must be cooled or water vapor must be added until the air is supersaturated. This means that it must contain more water vapor per unit volume than air at that temperature and pressure would contain if it were in direct contact with water or ice. Secondly, the air must contain particulate matter to enable the phase transition from vapor to water or ice to take place. Particles on which small water droplets or ice crystals form are termed **condensation nuclei.** Air which is entirely free of any particulate matter might become very highly supersaturated before any condensation occurs. Water droplets or ice crystals form best on particles with diameters from about 0.05 to 1 micron. The number of particulate nuclei available is important in determining the concentration of cloud droplets. Typically there may be about 3 million per cubic foot in oceanic air, and more in air over land masses.

Because the droplets or ice crystals formed on the particles are very small, there is a natural tendency toward further aggregation to form larger drops or ice crystals. This aggregation step is necessary for precipitation to occur. To fall out of the cloud, droplets or crystals must grow larger, either by adding more water from the water vapor, or by aggregating. The processes leading to formation of aggregates large enough to form precipitation are complex, and not entirely understood. Circulation within the air mass of the cloud is important, as is the nature of the condensation nuclei available. The effectiveness of a particle in causing nucleation is related to its "wettability," that is, its ability to make effective contact with water. Particles composed of substances which have the greatest affinity for water, such as salts of various composition, are most effective. The ability of a particle to induce precipitation is also a function of size. While large numbers of particles 1 micron and smaller are available to act as condensation nuclei, somewhat larger particles are more effective in promoting formation of the larger droplets or crystals necessary if precipitation is to occur.

Although the quantities of aerosol required to effect precipitation are very small, a surprising amount of solid material falls from the sky in rain and snow. For example, precipitation in North Carolina has been found to contain about 2.3 parts per million of sulfate salts. Assuming about 40 inches of rainfall per year, this amounts to about 7 tons of sulfate per square mile. In upstate New York the quantities of sulfate are twice this high. The sulfate probably arises largely through addition of oxides of sulfur to the atmosphere in burning of fuels, notably coal. Sodium chloride, which is present in similar concentrations in rain and snow falling on coastal states and over water, gets into the atmosphere mainly by evaporation of ocean spray, as mentioned earlier.

Modification of cloud formation and behavior by seeding the atmosphere with tiny crystals of a salt has been the subject of considerable research. Most commonly, an aerosol of silver iodide crystals, which are thought to be especially effective in bringing about nucleation and aggregation, is put into the air at some point in the atmosphere judged to be propititious for cloud formation and precipitation. The efficacy of this technique in inducing precipitation is open to question.[1] The potential benefits of a successful procedure for controlling precipitation to even a limited degree are sufficiently great, however, to encourage continued search for more effective techniques.

Since minute solid aerosol particles play an important role in initiating cloud formation, they are an important element in determining the earth's radiation balance. We have seen that the planetary albedo is related to the extent of cloud cover (page 23). An increased concentration of airborne particular matter might, if it were effective in initiating cloud formation, result in an increase in the albedo, and thus in a cooler climate.

Solid aerosol particles in the atmosphere may also affect the earth's albedo in their own right. They scatter the incoming solar radiation very strongly; some of the scattered light is returned to space. The idea that airborne particles are important in affecting the climate has formed the basis of a theory of climate change applicable to the past history of the earth and possibly also to its future. Huge volcanic eruptions of a type which would place large quantities of very tiny particles in the stratosphere and upper troposphere could place a pall over the globe which could have resulted in a sharp increase in the albedo for a protracted period, perhaps many years. The scattering effect of the aerosol could have been reinforced by the increased cloud cover which might have resulted from the higher concentration of nucleating particles.[2] Together these effects could bring on wintry periods of extensive glaciation.

There is considerable experimental evidence that volcanic

eruptions do have an effect over large distances on the solar radiation received at the surface. The most spectacular event of this type within the past century, and thus for which measurements of any value are available, was the eruption of Krakatoa in Java in 1883. The only solar observatory in existence at that time was at Montpellier, France. It required three months for the ash of Krakatoa's eruption to move to western Europe, and to decrease the solar radiation. The record of continuous observations before, during and after this time shows that the solar radiation averaged nearly ten percent below normal for three years. According to estimates based on optical effects made at the time, the average particle size of the Krakatoa ash was somewhat larger than 1 micron. The unusually high concentrations of particles capable of scattering visible light caused spectacular sunsets.

In more recent times, the eruption of Katmia in the Aleutian Islands in 1912 caused observable decreases in solar radiation intensity throughout the northern hemisphere. These and other observations relating to decreased solar radiation following volcanic activity support the theory that major climatic changes in the past may have occurred as a result of heavy volcanic activity. They also serve to sustain the hypothesis that atmospheric turbidity, arising from whatever cause, is an important variable in determing world-wide climatic conditions. There seems little doubt that man's activities are contributing to increase atmospheric "dustiness."[3] Measurements at Mauna Loa observatory in Hawaii, far from sources of pollution, suggest an increase of about 30 percent per decade. Similar measurements in Europe and elsewhere leave little doubt that the atmosphere is growing less transparent to the sun's rays, despite the fact that there has been no significant volcanic activity in the past few decades. It has been suggested that this increasing turbidity is responsible for the slight decrease in the average surface temperature of the earth during the past thirty years. This might be occurring because of increased scattering of light by the particles and because of increased cloudiness resulting from the higher concentration of nucleating particles.

The effects of increasing concentrations of particulate matter in the atmosphere are felt most strongly over urban areas. The degree to which the solar radiation is diminished in an urban area as compared with a nearby rural area is quite remarkable.[4] For example, a Russian study showed that Leningrad in the winter received up to 70% less sunlight than the surrounding countryside. In the summer the loss as compared with the countryside was less severe, about 10%. The heavier loss in solar radiation in winter is only partly due to higher concentrations of particles during this time. The angle which the sun's rays make with the surface is also important. Ultraviolet radiation is particularly strongly attenuated when the sun's elevation is lower, as it is in winter. As a rough overall estimate, assuming a constant level of particulate matter, cities receive about 15 to 20 percent less insolation than their rural environs. Aside from the objectionable effect of this screening of the sun's rays on the quality of life in the cities, it affects the climate by cutting off some of the heat input to the urban complex.

Assuming, in the absence of evidence to the contrary, that particulate matter which increases the atmospheric turbidity is also important in affecting world-wide climate, there is still a question as to which additions induced by man's activities are most important. It appears that increased dustiness of land surface cleared for agricultural purposes and in lumbering makes some contribution. This might be brought under control by improved land management practices, and systematic large scale attempts to reclaim areas which needlessly have become arid. The second major source of aerosols is industrial effluent. In the more heavily industrialized nations the smokes from combustion processes and metallurgical operations such as steel-making and ore roasting are responsible for a large efflux, much of which remains in the atmosphere for a long time. The need for cleanup of these sources is now being recognized in the United States. It may be hoped that the increasing additions of particulate matter to the atmosphere can be abated during the next decade and eventually stopped altogether. Aside from the climatic implications of a reduction in atmospheric particulate matter, there are substantial health benefits to be gained. Many of the particles which we breathe in during our lives are known to be harmful, especially in causing lung diseases, including lung cancer. Extensive evidence based on epidemiological studies suggests that the incidence of respiratory illness is directly related to the level and type of particulate matter in the atmosphere.[5]

There has been considerable speculation, and more than a little concern, regarding the possible climatic effects of the supersonic transport (SST) planes now being planned. The United States version of the SST, until recently under contract to the Boeing Company, is a four-engine aircraft 298 feet in length, and accommodating up to 298 passengers.[6] It is designed to fly at an altitude of about 12 miles during the supersonic stages of flights.

The SST program has been subjected to heavy criticism because of the undersirable effects of sonic boom resulting from flight at supersonic speeds.[7] In response, the Federal Aviation Authority has indicated that supersonic flight speeds would be permitted only over the oceans.

A second source of objection to the SST has been fear of its adverse effects on climate. The planes will be flying in the stratosphere for much of their flight times. Combution of fuel will result in the additions of water vapor, carbon dioxide and particulate matter at an altitude of about 12 miles. As a basis for estimation of the quantities involved, it is commonly assumed that there will eventually be a world-wide fleet of about 400 SST planes, each flying about four flights per day. The amount of carbon dioxide added to the stratosphere by this amount of flying will not add significantly to that already present. On the other hand, the amount of water vapor added is relatively much greater. Recall that water vapor levels in the stratosphere are very low, on the order of three or so parts per million. Water vapor tends to remain in the stratosphere for quite long periods of time, on the order of about two years.[8] Most of this water vapor enters the stratosphere in the tropics, where upward air motions are most pronounced, and returns to the troposphere in the middle latitudes. The rate of deposition of water vapor in the lower stratosphere by the four hundred SST planes will be about equal to the total rate at which water vapor is being added by atmospheric circulation. It is reasonable, therefore, to assume that there might be up to twice as much water vapor in portions of the lower stratosphere as at present.[9]

The humidity in the stratosphere is not particularly constant. Recent experiments conducted with stratospheric balloon flights have indicated that the water vapor level there has increased from about two parts per million to three parts per million during the past six years.[10] There is some question, therefore, whether a significant increase resulting from SST flights could be of climatic importance. Because the stratosphere is so low in temperature, only very low concentrations of water vapor are required to saturate the air. However, the low pressures which prevail make it difficult for the water vapor to nucleate, forming ice crystals. Quite possibly, then, the most important effect of the SST will be to add submicron particulate matter to the lower stratosphere, promoting the formation of characteristic high altitude ice crystal clouds. The effects of such cloud formation on climate is not clear. In a study conducted some years ago Haurwitz [11] estimated that high altitude cirrus clouds have a reflectivity of about 20 percent toward solar radiation, whereas low altitude clouds have reflectivities in the range of 70 percent. High altitude clouds, therefore, do not increase the albedo as much as lower altitude, denser clouds. At the same time, they are fairly effective absorbers of the upward long wavelength radiation from earth. If additional cloud cover produced by the SST flights should indeed prove to have such properties, it will contribute to an increase in the earth's temperature. Considerations based on the Manabe-Wetherald calculations [12] suggest, however, that temperature effects will be small, perhaps on the order of a 1°F increase in the average surface temperature. On the other hand, particulate exhaust would in itself contribute to an increase in albedo, and thus to a decrease in the planetary surface temperature. Obviously, suppositions about the long term climatic effects of the SST are largely conjectural. In any case, it seems quite likely that from about 1980 onwards, supposing that the SSTs are flying, the skies will cease to be as blue as we now know them. Perhaps future generations will learn to accept a hazier sky of diminished blueness. Many people are of the opinion, however, that the SST program is a commonplace engineering enterprise of questionable social and economic value and not worth the slightest sacrifice of any environmental value.

As with so many technological innovations, the *modus operandi* in the SST programs has been to assume that all will be well and to base program scheduling on economic and logistic considerations. The absence of program steps involving sustained testing of only a few aircraft, coupled with careful observations to determine possible environmental effects, can be related to the fact that SST development has been another of those more or less irrational international competitions which brook no slowdowns for such ethereal considerations as the long-term welfare of humanity. As representative E. P. Boland, Chairman of the House Sub-committee on Transportation Appropriations said in 1970, "The real question is whether we're going to lose the magnificent lead we have in the production and sale of aircraft around the world." The recent defeat of the SST appropriation bill in the United States Congress is an encouraging sign of increased governmental responsiveness to public concern for the environment.

6 coming up on STOP

The material progress of human society is measurable by its per capita energy consumption. An adult subsisting on a diet of about 2,500 kilogram-calories * per day consumes energy in simply maintaining his bodily functions at about the same rate as a 100-watt light bulb. This energy is eventually all transferred to the environment as heat. In a primitive society which had access to no significant additional sources of energy, this 100 watts would represent nearly all the per capita consumption of energy. In a more advanced society which is not highly industrialized, a certain amount of energy leverage is available through the use of water power, and fire for cooking, heating, metallurgy and other uses. This added leverage raises the per capita rate of energy consumption to something like 200 watts. In the United States, materially the most advanced nation on earth, the present rate of energy consumption is about 10,000 watts per person. This energy is generated by many different means, and is used in providing the various modes of transportation, in heating and air conditioning of homes, offices, stores and factories, and in the operation of industrial processes in the economy.

At the turn of the century the per capita energy consumption was only about 15 percent as large as the 1970 value. We have not arrived at a plateau on the growth curve of energy use. It is estimated that per capita energy consumption

* These are the units in which food "calories" are measured.

will continue to increase at the present rate of 2.5 percent per year for many years to come.

The rest of the world lags behind the United States in energy use. The 215 million Americans, constituting about 7 percent of the world population, are responsible for about 34 percent of the world's energy consumption. On the other hand, the rate of increase in world-wide energy consumption is more rapid than for the United States alone. If there is no devastating thermonuclear war, and if the human population can be stabilized at a level consistent with the spatial and nutritional resources of the planet, the growth in energy consumption in the less highly technological societies should be quite rapid during the next 50 to 100 years. Despite man's many potentialities for self-destruction and decline, we must assume that there is a civilized future in store for the human race, and anticipate the problems which that future may bring. Large quantities of energy are essential to the development of a technologically sophisticated world civilization, but there are limits on its use. Recognition of these limits and the invention of social mechanisms for the control of energy consumption are essential if we are to avoid at least one particular avenue for humanity's decline.

World resources which might be drawn upon in deriving needed energy have been very well discussed by M. King Hubbert in the book *Resources and Man*.[1] We might consider first the natural sources of energy, i.e., processes which occur in nature, into which man might intrude at some point to extract energy. If it were possible to convert the solar radiation falling on the earth's surface directly into other useful forms of energy—for example, into electricity—the sun itself might serve as the supplier of our energy needs. While it is certainly possible to effect this conversion, there does not appear to be much prospect at present of utilizing solar energy to generate electricity on a large scale, or for home heating, etc. Since the sun shines on a given place only part of the time, and with variable intensity, it is necessary to make provision for storage of excess energy for later use. To collect enough energy to produce as much electricity as a modern steam power plant would require covering an area of 16 square miles with collecting devices, as well as facilities for conversion to electricity and storage.

Hydroelectric power has been a significant factor in the development of electric power generation in the United States and Russia. Utilization of all feasible sources of stream flow in the United States would make possible a generating capacity considerably higher than is installed at present. It has been estimated that world-wide utilization of all practicable sources of hydroelectric power would make possible the generation of about *twice* the total electrical power now being generated throughout the world.[1,2] Three factors severely limit the prospects for water power as an answer to the world's future energy needs. First, utilization of all the feasible hydroelectric power sources would require sacrifice of considerable land area and would involve destruction of much beautiful scenery and wild land. Secondly, the impoundment of water by large dams may have adverse ecological effects on the areas associated with the impounded waters, resulting in decreased agricultural productivity or unfavorable climatic change. Finally, the basins created by the impoundment tend to collect silt and gradually lose their effectiveness as a storage for electricity generation. It therefore appears that water power will contribute only a fraction of the power needs of the future.

The rise and fall of the oceanic tides causes the water level in certain partially enclosed basins to rise and fall many feet each day. For example, at Mont Saint-Michel on the French coast, the average range of the water level is 29 feet, and the basin in and out of which the tides flow has an area of about 240 square miles. The flow of water in and out of such a bay may be partially converted into electrical energy. A few tidal energy electric generating plants have been successfully built and operated. The amount of electrical generating capacity which might be realized by maximum practicable utilization of this source is, however, only a small fraction of the world's power needs. Similarly, tapping the energy of superheated steam escaping from the earth in areas of volcanic activity cannot be counted upon for more than what would be a small contribution in relation to the total requirement.

In summary, among all the natural sources of energy, only hydroelectric power can make a sizable contribution to future energy needs, and only at a cost in terms of deterioration of the environment which will surely grow unacceptably high as land resources increase in value. We must, therefore, look to man-made sources of energy to satisfy the bulk of present and future needs.

The world resources of fossil fuels and the expectations for their future consumption were described in Chapter 4. To recapitulate, the supplies of oil and natural gas will be largely exhausted within another 80–100 years, and about half the estimated recoverable coal reserves will be consumed before 2100. The forecast for natural gas and oil consumption seems firmer than for coal. Consumption rates for the latter will be limited to a large extent by increasing use of nuclear energy for electric power generation, but increased on the other hand by increased use of centrally generated electric power for satisfaction of energy needs.

Nuclear fission provides the energy employed in nuclear power plants. The energy is available in the form of heat from the nuclear reaction. The heat energy is converted into electricity by heating water to steam, superheating the steam, and then passing it through a turbine which drives an electric generator. Nuclear power plants differ from power plants which employ coal, oil or gas only in the process used to heat the working substance, water.

The nuclear power plants operating today are of a type which consumes a particular form of uranium, uranium-235. This is a relatively rare isotope (form) of uranium; it represents only 1/141 of natural uranium. To extend the world's resources of fissionable materials, it will be necessary to convert to so-called breeder reactors, in which normally non-active forms of both uranium and thorium can be converted into fissionable forms, while at the same time heat is being produced for use in generation of electric power. Assuming that satisfactory breeder reactors can be developed in the next twenty years, the supply of potentially fissionable materials is very large. There should be enough raw materials to provide energy for man for an almost indefinite future. It is very important, however, as Hubbert emphasizes, that the breeder reactor program be developed with all possible speed. If the supply of the rare uranium-235 should be exhausted before adequate breeder capacity is developed, there will be no way in which the inactive uranium and thorium can be activated on a sufficiently large scale.

Nuclear fusion, which forms the basis of the hydrogen bomb, has not yet been controlled. While it is possible to initiate the reaction which produces the dreadful explosion of a hydrogen bomb, there is as yet no known way to control the reaction to obtain usable power over a period of time or on a continuing basis. A way may never be found. Certainly it cannot be said with confidence today that the enormous technical difficulties will be overcome in the forseeable future. Assuming, however, that eventually the necessary science and technology should be forthcoming, and that a controlled fusion reaction could be made to provide energy for generation of electricity or could in some other manner be brought to useful form, man would have at his disposal an essentially limitless supply of energy. The nuclear fusion reaction which is potentially the most useful involves deuterium, a form (isotope) of hydrogen which is present in all hydrogen-containing compounds, including the waters of the oceans, to the extent of one deuterium atom for each 6,500 hydrogen atoms.

This very brief survey of the prospective energy supplies serves to make the point that, in terms of total energy needs, man need not anticipate a shortage. There are problems associated with the utilization of each source of energy. We saw in Chapter 4 that burning of fossil fuels leads to an increase in the carbon dioxide content of the atmosphere, with probably serious consequences in terms of climatic change should the consumption of fossil fuels be carried too far. On the other hand, nuclear power plants must cope with highly radio-active wastes, and must anticipate possibly calamitous consequences in the event of a major accident.[3] There is reason to expect, however, that with development of a thoughtful, long-range program for radio-active waste disposal, and a diligently prosecuted inspection and control system in connection with plant operations, nuclear plants can generate power safely.[4]

There are differences in the economics of fossil fuel vs. nuclear plant operation which now determine which type of plant is built. Recent developments in coal gasification; improvements in transmission of electric power which make location of plants at the coal mines (mine-mouth operation) more attractive; the potential of magnetohydrodynamics (MHD), a new process for generation of electricity: these may keep coal economically competitive with nuclear power for many years to come. If, however, there are other consequences in terms of environmental effects of one form of energy use as opposed to the other, these should take precedence over purely economic considerations, which in any case are set more by governmental policies than by the forces of the free market place.

But if we can be satisfied that the human race will possess adequate energy resources for its future needs, there remains still another question. Are there any limitations on the *rate* at which energy is produced, in terms of environmental effects? All forms of energy generation, and all uses of energy in its many forms, ultimately lead to the production of heat. For example, electrical energy, once produced, may be used for any number of things. Obviously, if it is used to operate an electric stove or heat a home, it is re-converted directly back into heat. However, even if it is used to operate a television set, the compressor of an air conditioner, a dishwasher or an electric automobile, essentially all of the electrical energy which goes into the operation still ends up as heat.

As we have seen, the earth's climate is the product of a nice balance of heat flows, based largely on the heat input from solar radiation. If the intensity of this heat input were to vary, the average surface temperature of the planet would change also. It has been estimated that a 1 percent change in the insolation (solar input) would produce 2.5°F change in the average surface temperature.[5] But now suppose that there were a change in the heat at the earth's surface, not because of a change in solar intensity, but as a result of man's energy consumption. The effect would be largely the same. A major difference in the two occurrences is that a change in the insolation would be uniformly distributed, whereas energy generation by man is restricted largely to land areas, and is very highly concentrated in certain regions. These "heat islands" will experience effects of energy generation more seriously than other regions. Over a period of time, however, the movements of air masses should result in a distribution of much of this added heat energy, so that the world-wide climate might be affected if the quantities were large enough.

The current rate of world-wide energy consumption is 5.5×10^9 tons of coal equivalent. In heat terms this comes to the huge total of 3.7×10^{16} kilogram-calories. In terms of the energy consumption rates discussed earlier, it amounts to about 1500 watts per person, averaged over the earth's population of more than 3 billion. If we assume that this energy is evenly distributed over the earth's surface and over the course of a year, it amounts to 1/25,000 of the averaged total insolation, that is, of all of the solar radiation not reflected back into space. This is a very small fraction, but it must be remembered that energy consumption is in fact not evenly distributed over the earth's surface; most of it is confined to the land areas of the Northern hemisphere. If the earth's population were to stabilize at about twice the present level, and if it were to enjoy the same per capita rate of energy consumption which now prevails in the United States, about 10,000 watts, the total energy consumption of the human race, averaged over all the surface of the earth, would then be about 0.05 percent of the total insolation. The planet can easily accommodate this level of energy generation if it can be controlled to avoid overlarge local concentrations. But how much can it accommodate without adverse effect on the planetary climate? Even assuming the most optimistic projections for our technical capabilities in the future, it does not seem likely that the overall rate of energy generation could be more than about 0.5 percent of the insolation without producing adverse environmental effects.[6] Even this level, which amounts to 6×10^{14} watts, would require very substantial advances over present technology in energy transmission and heat distribution. Unless much of the waste heat could be released into the oceans the land areas would experience excessive heating. To prevent excessive localized heating it will be necessary to locate sources of electricity generation close to or, more probably, in the oceans. To transmit the enormous amounts of required energy to inland centers it will be necessary to build and maintain underground cable systems of superconducting materials, capable of carrying tremendous currents. At present only a relatively few substances which can act as superconductors are known, and these do so only when maintained at extremely low temperatures, near absolute zero (–459°F). Maintenance of the required low temperatures along cables hundreds of miles long is beyond present technology. A great deal of research and development will be required before we are in a position not only to build and operate huge nuclear-fueled power plants over or in deep ocean water, but also to arrange for transfer of the electricity so generated over great distances to the points of use.

The estimated upper limit of energy consumption will be of only trivial interest if human society cannot find a way to stabilize population at a level which permits application of technology to improving man's material lot. The prospect of thermonuclear warfare grows more substantial with the continuing failure of national governments to join in meaningful arms control.[7] The general annihilation resulting from thermonuclear war would make any concerns about overuse of energy irrelevant for some time, and possibly forever. But if humanity can, like Leopold Bloom in Joyce's *Ulysses*, somehow manage to outwit the denizens of Barney Kiernan's tavern, this upper limit will be a serious factor in planning for the future beyond about 2050. Suppose that the per capita consumption of energy in the United States continues to grow at the present rate, with a doubling time of less than 40 years. Then in the year 2050, if there are only (!) 7 billion people on earth, and if they have the same per capita rate of power consumption as the people of the United States at that time, the total energy consumption will be more than 0.2 percent of the solar influx averaged over the entire planet. Another doubling in per capita consumption beyond that would put the total at the estimated upper limit.

These figures are of interest because they show that man is not so very far away from certain fundamental limits on the growth potential of a factor which has been traditionally assumed to be available in limitless quantity. It would be the grossest folly to ignore the implications of these limits in planning the future. The control of energy consumption on a world-wide basis will require a degree of cooperation between peoples which is probably unattainable with the nationalistic structure of present world society.

Aside from these questions, which are long range only in a relative sense, there are more immediate problems posed by generation of energy in highly developed societies. Figure 6–1 shows projected total energy demand in the United States in various categories, as developed by R. T. Jaske and associates.[8] These data show that the total energy dissipation in the year 2000 will be about three times as great as at present. The total of about 6.7×10^{12} watts converts to about 0.37 percent of the insolation, averaged over the coterminous land area of the United States (that is, excluding Alaska and Hawaii). Still more significant is the fact that the generation of this energy is not evenly distributed. For example, the eastern seaboard metropolitan area, extending from Boston to Washington, D.C., encompasses a heavy concentration of energy demand. Assuming that the per capita energy consumption in this region is the same as for the nation as a whole, it follows from straightforward calculations that the total man-made heat dissipation in this region in 2000 will be on the order of 15 to 50 percent of the ground-level insolation in the region, depending on the time of year. This is an enormous burden on the atmospheric environment, which must accept this heat in the form of direct heating or through the transfer of latent heat to the air by evaporation of water. In either case, very pronounced regional climatic changes may be expected. The air over the remainder of the nation also will experience thermal loading in varying degrees. Centers of high population density, such as the Chicago-Detroit-Cleveland chain, will need to cope with similar heat problems.

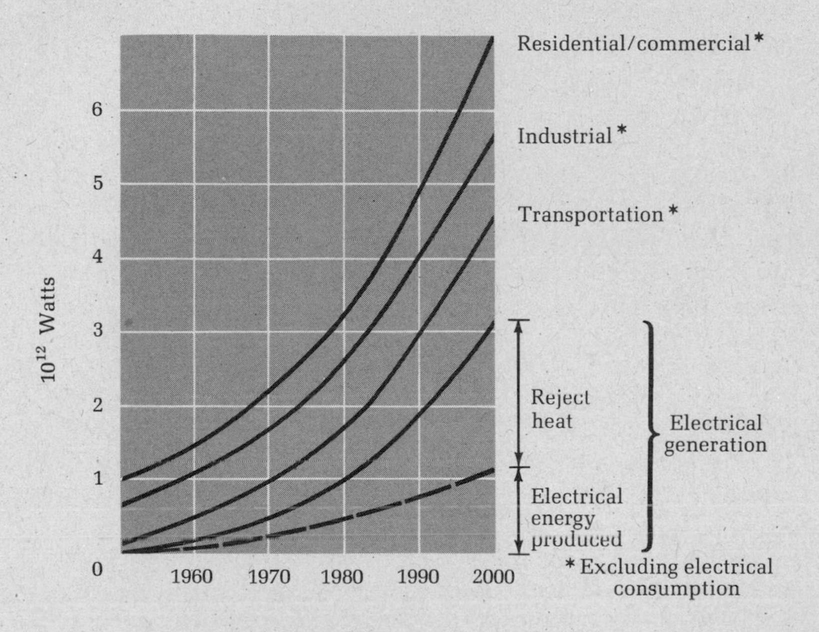

FIGURE 6–1. *Projected energy and waste heat production in the United States.*

The most obvious climatic change resulting from regional concentrations of heat dissipation is that the atmosphere will be warmer in these regions; how much warmer depends on the overall weather situation. Heating effects will be most evident when there are no large inter-regional movements of air. In these circumstances there may be strong convective forces in the centers of the heat islands, producing cloud cover and rain.[9] A great deal will depend on the extent to which the water content of the heated atmosphere has been increased by evaporative cooling of water supplies. During heat waves, when the demand on the energy supplies may be particularly high, localization of heat within the affected region could produce exceptionally high temperatures, resulting in still further demand on the power supply system.

Areas adjacent to the large heat islands might experience much increased cloud cover, rainfall and incidence of fog. A number of other factors, such as atmospheric turbidity, will be important in determining the details of the climatic changes which occur. Although much remains to be learned about the climatic effects of localized heat dissipation, it is evident, on the basis of urban climate studies which have already been carried out, that regions of high density energy consumption will experience increasingly altered climates in the years ahead. Thermal pollution is normally considered to refer to the undesirable consequences of adding heat to water bodies. Because of the intimate coupling between water and the atmosphere, however, the notion of thermal pollution must be more broadly conceived to include the addition of heat to the total *environment* in quantities which adversely affect the quality of life.

7 thermal effects of power stations

The prospective difficulties in large scale heat dissipation described in the previous chapter are presaged by the difficulties in disposal of heat presently being experienced at local sources.[1-3] All energy generation, no matter what form it may take for the moment, eventually appears as a heat burden on the environment.* Because electricity is assuming a continuously increasing share of the energy market, and because the problems of heat dissipation from large power plants are particularly acute, the impact of the rapidly growing electric power industry on the environment has already become quite evident, and has aroused wide concern and government attention. Whereas total energy use in the United States is increasing at a rate of 3.3 percent annually, electrical power generation is increasing at about 8.8 percent annually. The percentage increase is quite well distributed throughout the nation. This means that regions that already have a high concentration of electric generation capacity are getting the major portion of the increased power installation. By 2000, electric power consumption and the associated waste heat will account for nearly half the total energy dissipation in the United States (see Figure 6-1, page 69).

---

* A small fraction of industrial energy is devoted to processes which store energy in some other form. For example, the electrochemical process used to produce aluminum results in conversion of electrical energy to chemical energy. Even so, a large proportion of the total electrical power consumption of the aluminum industry is dissipated as heat.

Figure 7–1 shows how the total electric generating capacity is expected to grow to 2020, and shows also the projected apportionment between nuclear, fossil fuel and other sources. It should be noted that this figure illustrates the division of just the lower section of Figure 6–1, and does not include the waste heat part of the total energy dissipation due to electricity generation.

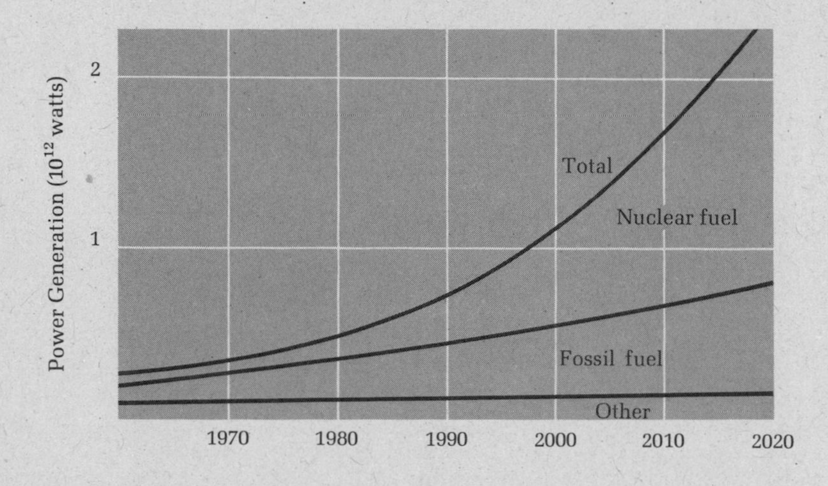

FIGURE 7–1. *Projected Apportionment of Electrical Generating capacity to 2020.*

Although in many situations electrical energy can be utilized more effectively than other forms, there is a considerable heat waste in the generation of electrical power. The steam-electric plant is the major means of conversion of fossil fuel or nuclear energy into electricity. A schematic diagram of such a plant is shown in Figure 7–2. Water is converted into steam in a boiler which is heated either by burning fossil fuel or by a nuclear reaction. The steam is superheated and then passed at very high velocity into a turbine. The energy stored in the steam is converted into mechanical energy, the turning of the turbine blades. The turbine shaft is attached to the shaft of an electric generator, where the mechanical energy is converted to electrical energy. The force which pushes the steam through the turbine derives from a tremendous drop in pressure, from about 2,000 pounds per square inch as it enters the turbine, to less than atmospheric pressure (which is 14 pounds per square

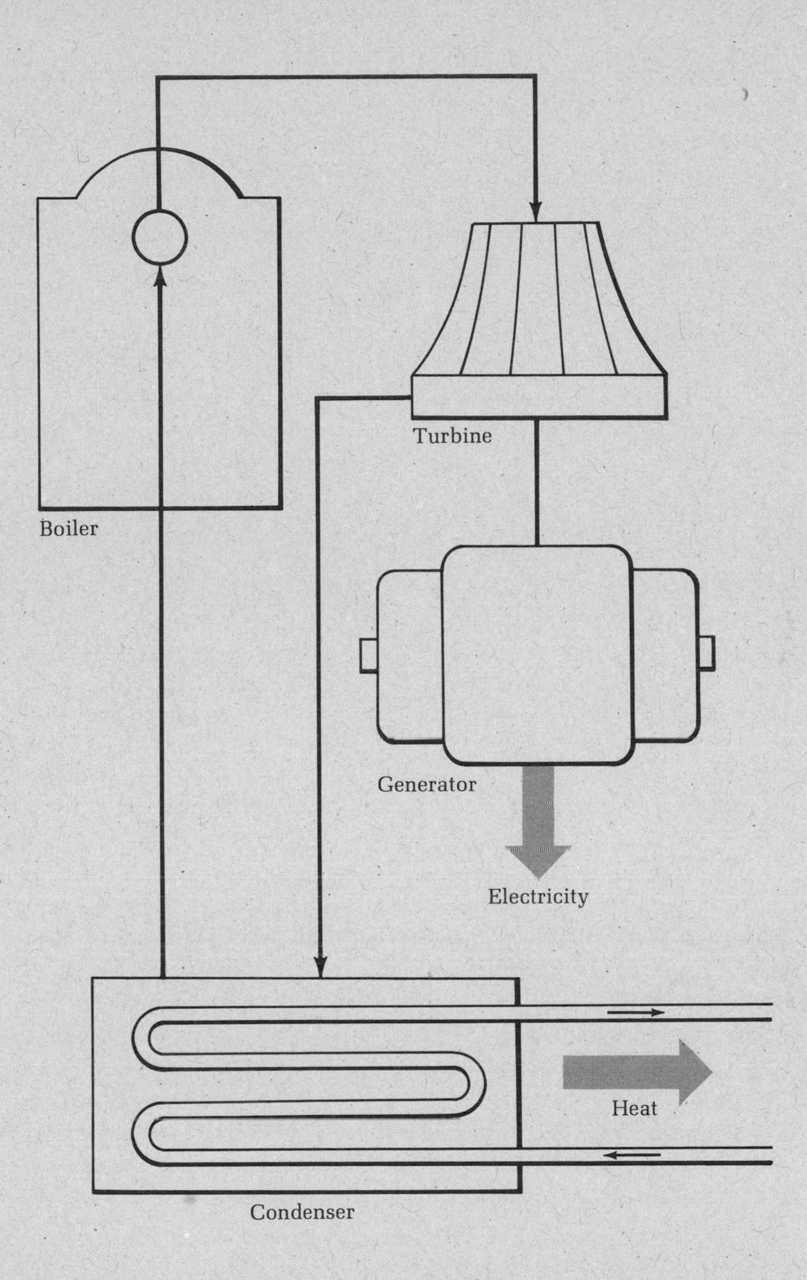

FIGURE 7–2. *Schematic diagram of a steam electric plant. The boiler is powered by burning of either nuclear or fossil fuel. Heat dissipation to the environment occurs primarily through the condenser.*

inch) at the outlet. This great pressure drop depends on having at the outlet of the turbine a condenser which cools the hot steam and condenses it back into water. The colder this condenser, the more efficient the operation of the plant can be.

There are certain limits on the convertability of heat into other forms of energy. These limits, expressed in the laws of thermodynamics, cannot be circumvented by clever engineering tricks; they represent the best we could do if everything went as well as possible. If there were no other losses at all in a typical steam plant, the maximum convertability of heat into electricity would be about 60 percent. In practice, there are other losses. Modern fossil fuel plants can operate with efficiencies as high as 40 percent. The current national average efficiency for plants of all ages and sizes is about 33 percent. This means that for every unit of electricity which comes out of an average power station, two units of heat *in addition to the electricity* must be rejected to the environment. Nuclear power plants, because of certain design limitations, cannot now attain efficiencies higher than about 33 percent; their waste heat requirements for a unit of electricity produced are therefore even higher than for a modern fossil fuel plant. It is anticipated, however, that the breeder reactor plants which will be coming into use in another 20-30 years will attain efficiencies of up to 44 percent.

A little of the waste heat from a fossil fuel plant is transferred to the environment through the hot gases emitted from the smokestack. A large fraction—85 percent—is transferred via the condenser which cools and condenses the hot steam. In a nuclear plant, which has no smokestack, essentially all of the waste heat is rejected to the surroundings via the cooling water. The condenser is a chamber filled with a large number of small diameter tubes through which cooling water is passed. The purpose in using a large number of small tubes rather than one large one is to increase the cooling surface area, and thus improve the transfer of heat to the cooling water. By passing a large quantity of cooling water through the tubes continuously, the temperature rise of the cooling fluid can be kept within the range of 10 to 30°F. The amount of heating which occurs depends on the steam generating capacity of the plant and the volume of cooling water circulated per unit time. For large, modern power stations, the volume of water required is very large.

In discussing the impact of power plants on water supplies and quality, it is helpful to distinguish between gross water usage and consumptive use. Any particular water user, such as a power plant or a city, requires a certain amount of water for its total uses, termed the gross use. It may, however, return much of this water to the body from which it was drawn. Hopefully the water returned is of a quality which does not result in degradation of the source. In any case, the difference between the amount withdrawn and that returned is the consumptive use. For example, a city may withdraw water from a river and return much of it through its sewerage treatment plant and storm drains. Its consumptive use represents water sprinkled on lawns and otherwise evaporated. A power plant has a consumptive use of water for makeup of water lost from the steam through leaks, the need for some flushing, etc. In addition, there may be consumptive loss resulting from the cooling function. The loss in this instance depends very much on the type of cooling system in use.

The gross use of cooling water by power plants at present is more than 50 trillion (50×10^{12}) gallons per year. It is expected that this will double by 1980. The total annual runoff of water in the United States excluding Alaska and Hawaii, that is, the total quantity of water which runs from the land into all the streams and rivers of the nation, is on the order of 450×10^{12} gallons. By the year 2000, if there were no change in the pattern of water use by power plants, the cooling water requirement would equal this total runoff. This comparison gives some idea of the magnitude of the overall cooling water requirement.

The size of a power plant is measured in terms of the capacity for generating electricity, usually expressed in millions of watts, or megawatts. Thus, a 100 megawatt plant is capable of generating electricity at the rate of 100,000,000 watts. Notice that a watt expresses the **rate** at which electrical energy is consumed. (For example, 1 watt corresponds to a rate of energy consumption of 14.3 calories per minute.) Thus a 100 watt lightbulb consumes electricity only 1/10 as rapidly as a 1000 watt (1 kilowatt) electric heater. The **quantity** of electricity used is given by the rate of use times the length of time it is being used. For example, a 100 watt light bulb burning for 10 hours uses up 1000 watt-hours, or 1 kilowatt-hour. The 1000 watt heater uses up 1 kilowatt-hour by being on for just one hour. Electricity is sold in terms of kilowatt-hours; in a typical home situation it costs about $0.04 per kilowatt hour. The rate scale varies with the quantity employed. The use of electricity in home heating is encouraged by providing a more favorable rate scale for those who have electric heating. Industrial users are similarly rewarded for using large quantities of electricity.

The sizes of power plants have been growing larger in recent years. The maximum size power plant in the United States has increased from about 200 megawatts 15 years ago to 1500 megawatts today, and even larger units are under construction. A typical large modern station has on the order of 1000 megawatts capacity. Much of the ensuing discussion is in terms of a plant of this size.

The simplest and most commonly employed method for provision of cooling water is the single-pass system. Water is withdrawn from an available source such as a river, lake, bay, estuary, or ocean, and pumped through the condenser. After passing through the condenser it is discharged back into the source. The points of intake and discharge are separated so that there is a minimal recirculation of water which has been through the system. The rate at which water must be pumped through the condenser depends on the size of the plant, and on the temperature increase which the water is expected to undergo. A 1000 megawatt plant would require about 500,000 gallons per minute if the temperature rise of the water is to be kept in the range of 20°F or so. This is equal to about half the total gross water usage of the city of Chicago.

After discharge the heated water mixes with the receiving body of water in what is termed a mixing zone. In most installations the mixing zone is not well-defined. For example, in a river installation, the discharge may be near the surface, and the warm water may spread out so that there is a continually changing temperature near the surface as one proceeds away from the discharge point. Water quality standards as set by the federal Water Quality Act of 1965 and by state regulations are for the most part based on temperature changes in the receiving body and, with one exception mentioned later, do not deal directly with the size of the mixing zone and the actual temperature rise which may occur in the water as it passes through the condenser. Typically, the water is heated from 10 to 30°F in passage. It is then discharged into the receiving body where it mixes with the water of that body in the ill-defined mixing zone. The overall effect on the temperature of the source body is determined by measurements outside the mixing zone. Once the heated water is discharged from the pumping system, heat is transferred to the atmosphere by direct transfer and evaporative cooling from the surface. The only consumptive use of water in the single-pass cooling system results from additional evaporation from the source water body as a result of the heat added. This may amount to about one percent of the gross water use, that is, of the condenser flow. The most efficient cooling of the heated discharge results when the temperature rise in the condenser is allowed to go fairly high, and the water discharge is kept near the surface, without a great deal of initial mixing of the hot water and the receiving body. In this situation, the temperature in the mixing zone is highest, and the rate of cooling by evaporation is highest. This, however, increases the thermal shock of organisms which pass through the condenser, and may result in temperature gradients in the receiving body which are lethal for some of the biotic species in that water.

It is often the case that streams of adequate flow rate are not available to provide the cooling water necessary for a large, modern installation. In this case a cooling pond or artificial lake may be employed. Depending on the depth of the water, and on climatic conditions, a water area of from one to two acres per megawatt is required. Thus, for a 1000 megawatt plant, a lake or pond with an area of 2 to 3 square miles is needed. The water is recirculated between the condenser and the pond. Cooling of the discharge water occurs through evaporative cooling, direct transfer of heat to the air, and radiation. The pond or lake must receive make-up water to compensate for the evaporation loss. If the lake is created by building a dam at a suitable site, there is usually enough runoff from surrounding land to maintain an adequate lake level. The cooling pond is generally designed to function purely as an adjunct of the power plant, and is seldom used for any other function. An artificial lake, if it is of adequate size, could also serve recreational uses. An installation of this sort requires a substantial capital investment in land purchase, lake clearing and dam construction. Typically it might be about two to three million dollars for a 1000 megawatt plant.

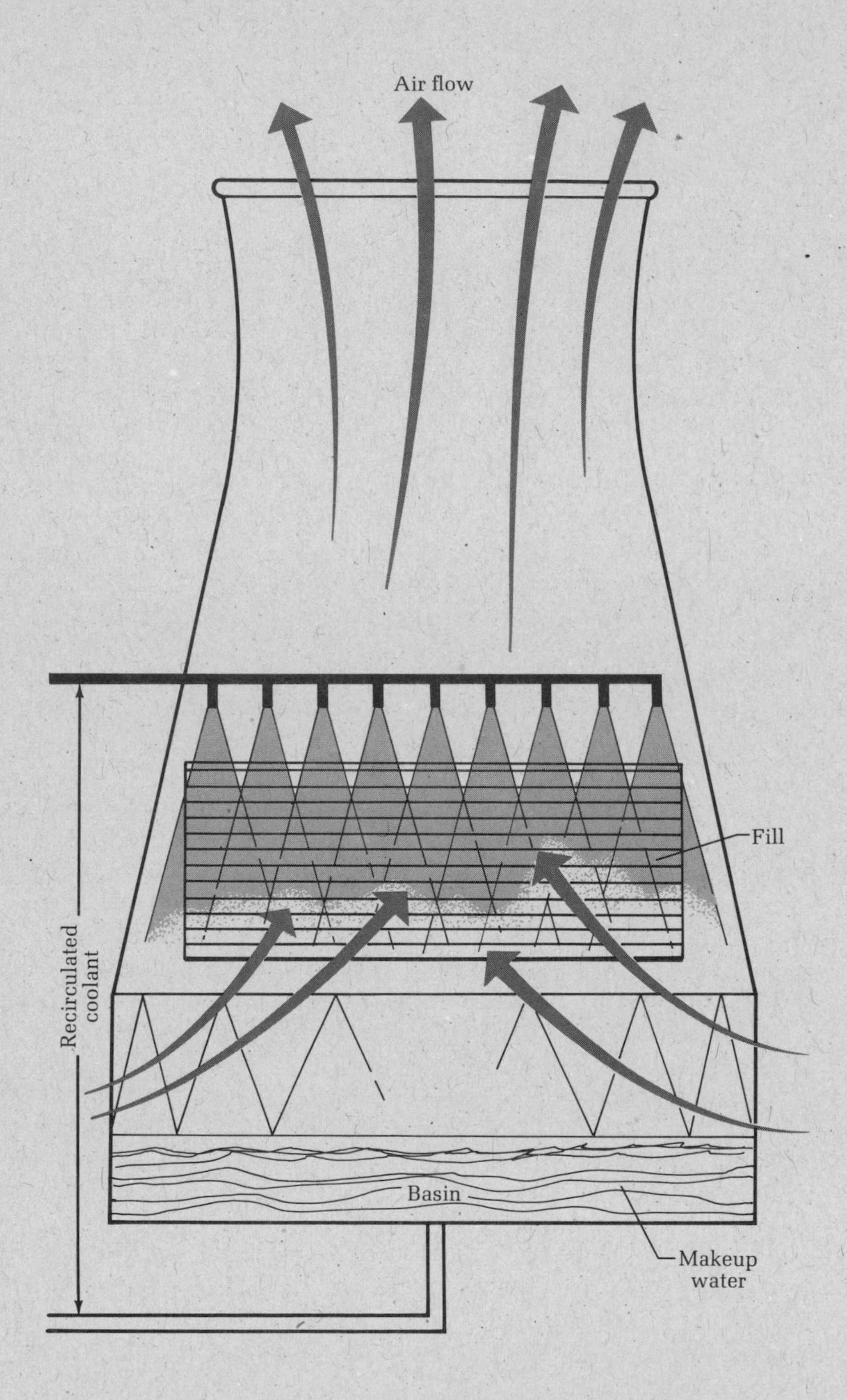

FIGURE 7-3. *Schematic diagram of a natural draft evaporative cooling tower.*

Where recourse to a cooling pond or artificial lake is not a viable course of action, the use of cooling towers may serve to increase cooling efficiency. The most common type of cooling tower is the so-called wet type, Figure 7–3. Water which has passed through the condenser is sprayed onto a wooden lattice network ("fill") inside a large tower which is open at the bottom. Figure 7–4 shows a wet tower installation which uses natural draft cooling. Evaporative cooling is promoted by movement of air over the water, which now has a large surface area.

FIGURE 7–4. *A natural draft wet cooling tower installation at the Tennessee Valley Authority steam plant, Paradise, Kentucky. Each tower is 437 feet high and is large enough at the base to comfortably hold a football field. (Photograph courtesy of the Tennessee Valley Authority.)*

The air may be moved by natural convective cooling, or by large fans in the bottoms of the towers. The cooled water collects in the basin at the bottom of the tower, and is re-circulated to the condenser. Makeup water must be added to replenish that lost by evaporation. The evaporative loss is about 3 percent of the flow rate. It is necessary to add chemicals to the water to retard the growth of algae and fungus on the fill, and to inhibit corrosion of pumps and other exposed metal. Since these chemicals and others present in the source water tend to accumulate, a periodic flushing of the system is required. If these washings are discharged to the source water body, they constitute a source of chemical pollution.

The cooling tower system embodies a very heavy emphasis on evaporative cooling as a means of disposing of the waste heat. Plants which employ this means have a much smaller gross use of water from available supplies, but a much higher consumptive use. A 1000 megawatt plant would require about the same consumptive use of water as a city of about 100,000 population. Under some atmospheric conditions the large quantities of water vapor added to the atmosphere may create problems with fog and icing in areas surrounding the cooling towers. Evaporative cooling is most effective when the relative humidity of the air is low. In these circumstances the air has a large capacity for adding water, and considerable cooling may occur. Natural draft towers may be appropriate under these conditions. Where the relative humidity is high, the potential evaporative capacity is lower, and mechanical draft may be necessary to effect the desired degree of cooling. It is in these circumstances also that fog and icing may occur in the area around the cooling installation. Because of their huge size, the initial capital costs are substantially higher for the natural draft towers than for mechanical draft systems. For a 1000 megawatt installation, the mechanical draft system would cost about $8 million as opposed to $12 million for the natural draft. Maintenance costs are, however, much lower for the natural draft systems.

In cases where addition of water to the atmosphere or the consumptive loss of water are unacceptable, a dry tower might be installed. The dry tower cooling system operates much like the cooling system of an automobile. Water is circulated through the condenser, and then into a giant radiator. Air is blown through by large fans, and the water is cooled by heat transfer through the walls of the radiator to the air. The air moving through the towers is heated considerably without the addition of any moisture, so its relative humidity is lowered. Since there is no evaporative loss, the consumptive use due to leaks and the need for periodic flushing is nominal. Contact between the circulating atmosphere is less direct than in a wet tower, so the movement of air must be much more vigorous. There is a loss in cooling efficiency as compared with evaporative cooling; the water can never be cooled to less than the ambient air temperature, whereas in evaporative cooling the evaporation process allows cooling below the ambient air temperature. (This is what makes the air feel so cool upon stepping from a swimming pool on a windy day.) As a result of these shortcomings, the dry tower system involves a 6 to 8 percent loss in overall output.

The largest plant for which a dry tower system has been installed is in Rugeley, England. The installation has a capacity of 120 megawatts; it employes one huge tower about the size of one of those shown in Figure 7-4. Despite what is written above about the need for fans, this installation uses natural draft. However, England has a cool climate, the power plant is not large, and the tower is very large. To provide a dry tower installation for a 1000 megawatt plant in the United States would certainly require mechanical draft with large capacity if the tower size were to be kept to anything reasonable. The capital costs for a 1000 megawatt installation are estimated to be about $28 million. The operating expenses would be substantial. High maintenance cost in addition to the operating expenses can be anticipated.

The dry tower has been heralded by some as the best means of minimizing the impact of the power plant on the environment. Representatives of the power industry, on the other hand, look upon the high capitalization cost with horror. There is the additional fact that a significant fraction of the power must be dedicated to operation of the cooling system. It is important to note, however, that these cost factors are not as important as they might at first seem. The cost of the 1000 megawatt plant itself, exclusive of the cooling system, is an estimated $120 to $150 million. If the costs of dry tower systems were apportioned uniformly to all users of the power, without discounting them unduly for the large users, they would result in an increase of only a few percent in the average homeowners' electric bill. What matters in the long run is the total cost to the consumer. Some of the cost is direct, and appears in his electric bill; some is indirect, and may appear in the prices he pays for other commodities. Even more indirectly, it may appear in the quality of his surroundings. For example, if, in connection with the use of cooling towers, there is a consumptive use of water for which the power company does not pay, the costs are nevertheless there in the form of increased water prices, environmental deterioration due to loss of stream flow, undesirable fog and icing from the cooling towers, and so forth. It is therefore important to know whether the dry tower system does in fact result in less environmental intrusion. Provided that waste disposal from the installation is controlled to prevent chemical pollution, as with any other industrial water user, there should be no significant impact of the plant on the local water system. However, the dry tower system maximizes the heating of air, and in such a way as to minimize the addition of water vapor. It is not at all clear what the consequences of this will be, but less rain, fog, and cloudiness in the downwind near environs would seem to be an obvious outcome. If the downwind area happens to be an urban complex already suffering at times from excess heat as a result of high energy consumption density, the additional heat load will not be welcome. There is also the question of the noise level generated by a huge array of blowers. There may be a large area of unacceptably high noise level around the plant.

Although the situation may change as installations employing cooling towers become more numerous, electric power plant operations have been perceived until now as possibly inimical to the quality of waters. It is not surprising that the question of what constitutes thermal pollution—that is, the undesirable as opposed to noneffective or even beneficial addition of heat to the environment—is surrounded with considerable controversy. The decision whether to install more elaborate cooling facilities than might have been planned for in the absence of regulatory pressures involves large sums of money in the form of initial capitalization and in maintenance costs. In each case the decision must be based on evaluation of what the long-term effect of a particular method of cooling will be on the ecology of the water system, taking into account all the other factors which impinge on that same system and affect the ecology in various ways.

Water bodies such as rivers, estuaries, bays and lakes differ a great deal in thermal structure, in the overall chemical content of the water, in the movement of water in and out, and in the nature and extent of plant and animal life. To give some idea of the factors which may be important in dealing with questions of thermal pollution, however, it is useful to refer to a model (Figure 7-5) which possesses the elements found in most inland lakes and reservoirs.

During winter months, especially in cold climates, a typical body of water becomes more or less isothermal, that is, the temperature is pretty much the same throughout, from top to bottom. During spring and summer, the upper layer of water is warmed by the sun. A certain amount of the upper water is mixed by wind action to form a layer of nearly constant temperature warm water called the **epilimnion.** In the lowest layer, the **hypolimnion,** the water is relatively much colder. Because colder water is denser than warm water, it remains on the bottom, and does not mix with the upper waters. In the interface of these two layers, the **thermocline,** the temperature changes rapidly with depth. Figure 7-5 also shows a typical temperature profile with depth. The temperature difference between the epilimnion and hypolimnion increases as the summer progresses and may be as high as 35 to 40°F. The relative amounts of water in the two regions vary with the time of year and with the particular body of water, depending, among other things, on the depth of the body with respect to its surface area. Typically, though, the epilimnion might extend to 30 or 40 feet in depth.

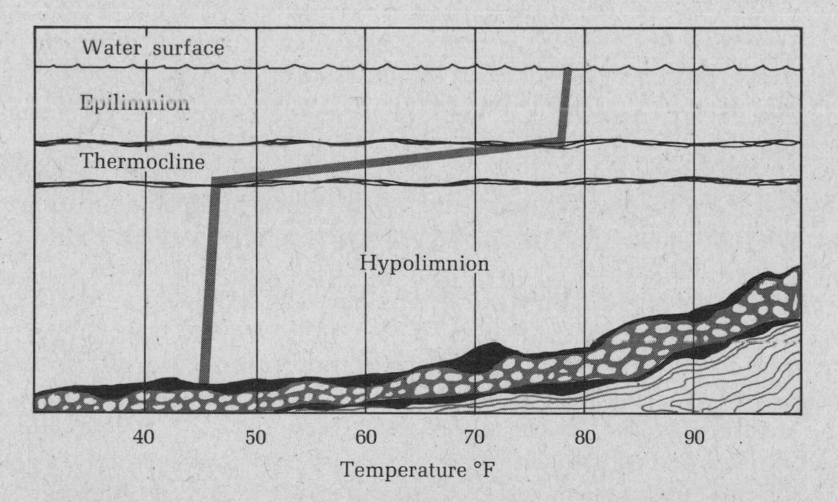

FIGURE 7-5. *Thermal stratification in a typical lake or reservoir during summer.*

The stratification just discussed is most pronounced in deep water bodies which do not have a large flow-through, such as a lake or reservoir. To a limited degree, however, the concepts apply also to rivers which are sufficiently deep and for which flows are not too turbulent. In estuaries, which are subject to twice-daily flushing by tidal action, such a thermal structure does not apply at all.

Cooling of the epilimnion in the fall causes the density of water in this level to increase. When it has cooled sufficiently, wind action produces a mixing of the epilimnion and hypolimnion. There is, therefore, a giant annual stirring action which mixes the entire lake contents, followed by a progressively increasing stratification before the cycle begins again.

During the warmer part of the cycle, biological production, particularly of plant material such as single-celled algae, occurs in the upper layer. It is here that the temperatures are sufficiently warm and that light is available for photosynthesis. Tiny plant organisms, called phytoplankton, and similar tiny animal organisms, called zooplankton, are formed in the epilimnion and live out their life cycle there. Oxygen for the epilimnion is continuously available because there is turnover of water in this layer and thus repeated exposure to the surface. Dead plant matter and planktonic organisms sink to the hypolimnion.

Animals which inhabit the hypolimnion, including many species of fish which require the colder water, must subsist on the oxygen dissoved in the water during the cold period of the annual cycle. There is very little diffusion of oxygen into the hypolimnion from above during the warm months when there is stratification. Oxygen in the hypolimnion is consumed not only by the fish which live there, but also by the bacterial decay of dead plant and animal matter which has sifted down from above.

Aquatic ecosystems are complex and variable, but it is helpful in discussing thermal pollution to know at least the major characters in the drama. These are outlined in Figure 7-6. The organisms along the bottom row provide in varying degrees the food for both invertebrates and fish. The latter, at the top of the food chain, are dependent on both invertebrates and plant materials for diet. Because they are at the top of the food chain, any alteration in the relationships between lower organisms is reflected in the numbers and kinds of fish which inhabit a water.

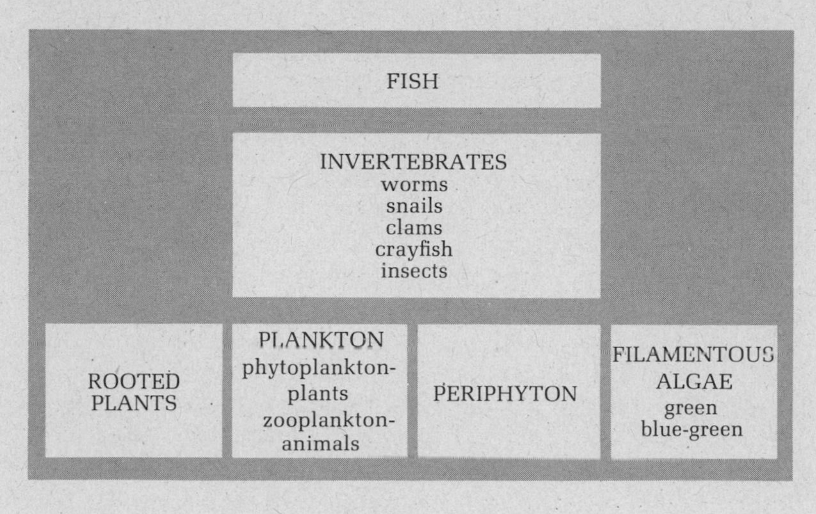

FIGURE 7-6. *Representative categories in an aquatic ecosystem.*

Most lakes gradually become more fertile with the passage of time. This process, known as **eutrophication,** results from the accumulation of organic wastes, and the addition of inorganic nutrients such as nitrogen and phosphorus from the surrounding watershed. With an increase in algal growth in the epilimnion, the water absorbs more solar radiation, thus warming this layer still further and stimulating still more biological production. Although eutrophication is a natural process for a lake, man has in some instances caused great accelerations in the process. Additions of nutrients such as nitrogen and phosphorus in sewage waste and in the form of runoff from lands heavily fertilized for agricultural use have provided the chemical base for accelerated plant growth. Thermal discharges may further accelerate the process by increasing the temperature of the epilimnion, extending the season during which it remains warm, or increasing the volume of this layer.

From the standpoint of power plant cooling efficiency, the best strategy is to remove cold water from the hypolimnion, employ it in cooling, and then discharge it into the epilimnion.

If there is a temperature difference of as much as 30°F between the two layers, a considerable heating of the circulating water can be effected, and water can still be discharged into the lake at a temperature no greater than that of the receiving body. The biological effects of this mode of operation are that the volume of epilimnion increases by the quantity of water heated in the power plant, and the volume of the hypolimnion correspondingly decreases. Since the **benthos** (organisms which live in the bottom level) depend on the dissolved oxygen in the hypolimnion, a decrease in the volume of this layer means a decreased capacity to support animal life. At the same time, the larger volume of the epilimnion and the increased growing season resulting from the addition of heated waters means that there will be more biological production, with a correspondingly greater demand on the oxygen in the lower layers for bacterial decay of plant and animal matter. Thus the fact that there is no increase in the temperatures of the various sectors of a lake's thermal structure is by no means an adequate indication that the thermal additions have no biological effects. The process of eutrophication in waters such as Lake Champlain, Cayuga Lake and—on a larger scale—Lake Michigan, can be accelerated by thermal additions which are physically arranged in such a manner as not to produce large temperature increases in the waters.

Eutrophication eventually results in a loss in oxygen available for fish such as trout that are adapted to the cold, dark, lower waters of a lake. With an increasing bloom of algae, the characteristic pea soup green color and texture of the water becomes increasingly prominent, impairing the quality of the water for swimming and water skiing. At a more advanced stage of eutrophication other recreational activities such as sailing, motorboating and fishing may be curtailed because of the presence of foul odors from decaying vegetation. Blue-green algae are often found in waters which reach an advanced stage of eutrophication. These plants are highly consumptive of oxygen, produce foul odors and tastes, and in some instances are toxic to fish, birds or mammals. In short, eutrophication does not lead to desirable changes in a lake. If fresh water lakes are to be preserved for the enjoyment of the present as well as succeeding generations, extensive efforts must be made to curtail additions of both nutrients and heat.

In addition to changes in the numbers and distribution of

biotic species which may result from eutrophication, other alterations may occur as a result of the thermal effects of power plant discharges. In rivers where the flow of water may not exceed by much the flow requirements of the condenser, the temperature of the entire water body may be increased significantly. During periods of warm weather, in which the ambient water temperature is already near an upper tolerable limit for many species of fish, added increases in temperature may result in the short run in occasional fish kills, and in the long run in elimination of certain species from the waters affected.

There is a basic idea in ecology that when an ecosystem is placed under stress, the changes which occur in the system result in simplification. Some species are eliminated, while other species which occupy a comparable position in the food chain and are tolerant to the stress simply take up the void created by loss of those which could not survive. This change is referred to as a reduction in species diversity. Unfortunately, nature has been rather perverse in that the plant and animal species which are most heat-tolerant are usually those we would rather do without. The undesirable blue-green algae are among the most heat-tolerant of plants. In a study of seven invertebrate species which are common to estuarine waters, the stinging sea nettle proved most hardy to temperature changes.

Discharges of heated effluent to rivers which host species of fish which migrate to spawn may produce a thermal block to the movements of such species unless care is taken to confine the area of temperature addition to only a portion of the river. The National Technical Advisory Committee to the Federal Water and Pollution Control Administration has established a number of temperature and mixing zone criteria applicable to a wide variety of water bodies. The water quality criteria on mixing zones suggest that 50 percent or more of the cross-sectional area or volume of flow of the stream or estuary be essentially free of temperature increase due to heat additions, to provide a pathway for fishes, and to ensure survival of free-floating and drifting insects, fish eggs, larvae and other organisms which may be temperature-sensitive.

Thermal effects of a power station may appear even though the temperature of the water body after mixing is not raised materially. Water which passes through the condenser is heated considerably, in a very short time interval. Some of the dissolved oxygen is lost from the water as it is heated because the solubility of oxygen in water decreases with increasing temperature. Free-floating organisms are of great importance in the aquatic life of estuarine waters. Studies have shown that phytoplankton, shellfish larvae, fish eggs and larvae, and invertebrate eggs and larvae, which are entrained in the water because they are small and free-floating, do not survive passage through the condenser very well. If this passage is accompanied also by a dose of chlorine, injected periodically as a biocide, their demise is virtually assured. Similarly, if passage through the condenser is followed by flow through a cooling tower, additional stresses there from added chemicals would virtually assure destruction of most of these organisms. Of course, if the use of cooling towers involves a closed or semi-closed system, with recirculation of most or all of the water, the impact on the water body is likely to be negligible.

The factors involved in an analysis of the effects of thermal loading of a particular aquatic system are very complex, and no single index such as change in insect population, aquatic plant growth or fish catch can provide an adequate testimony to the effects for good or ill of a particular installation. Sins of oversimplification are committed by conservationists who oppose certain installations as well as by defenders of the proposed sitings. It must be said, however, that the onus of misrepresentation rests more heavily on the electric power industry. Man does not have an auspicious record of having interfered with nature to good effect. Glowing descriptions of how much better the fishing is at the power plant outlet than in the surrounding water, given on more than one occasion in presentations before the Senate Sub-committee on Air and Water Pollution, are not an adequate response to the legitimate questions of scientists, conservationists, and others concerned with preservation of the aquatic regime.

8 too much is too much is pollution

Increased awareness of the potential damage to the environment from power plant operations has led to vigorous opposition to many new planned installations. In some cases this opposition has resulted in construction delays; in others the plans have been cancelled altogether. Although there are no really typical cases, examination of a few specific case histories serves to bring some of the issues involved into better focus.

Lake Michigan is a freshwater lake of considerable recreational importance, and the source of water for both a large population and a large industrial enterprise. It seems quite clear that Lake Michigan is eutrophying; the more pessimistic among the ecology-minded have already written it off. Whereas Lake Erie is clearly in an advanced stage of eutrophication, Lake Michigan, with a much larger volume and surface area, will take longer to reach a comparable condition. On the other hand, Lake Erie enjoys a rather substantial flow-through, whereas Lake Michigan does not. The successive accumulations of chemical pollutants of all kinds will not be flushed out of the lake for many decades.

The planned additions of many new power plants on the shores of Lake Michigan during the next ten to thirty years might accelerate the process of eutrophication if the additions of heat are significant in relation to the heat exchanges which naturally prevail between the lake and the atmosphere.[1] A few fairly basic considerations are helpful in forming some idea of the magnitudes of the various factors involved, and thus in deciding whether a serious problem is in the making. Most of the power stations now scheduled for completion during the next decade are to be located on the southern half of the lake. It seems appropriate, therefore, to assume that the lake area affected is about 12,000 square miles, rather than the entire lake area of 22,400 miles. By 1980 or shortly thereafter there will be about 10,000 megawatts of installed nuclear-powered generating capacity on the southern half of the lake. Assuming the plants are 33 percent efficient, and that open, single-pass cooling systems are employed, this means that there will be a heat release into the lake of about twice this, or 20,000 megawatts. Averaged over the assumed area of 12,000 square miles, this amounts to about 0.5 percent of the averaged insolation on the lake.

It can be anticipated that the water for cooling these plants will be taken from fairly shallow water, so the water will be part of the epilimnion, and that the discharge will be near the surface, or at least into the epilimnion. Assuming that this upper layer is an average of 50 feet in depth, the total volume for an area of 12,000 square miles is 1.67×10^{13} cubic feet. If all of the reject heat were retained in this upper layer of the lake, the temperature would increase about 0.6°F in a year. We can simplify the problem somewhat without changing its essentials by assuming that we have the lake at some reference condition typical of years before the addition of the new power plants. Now assume that all of the 10,000 megawatts of generating capacity is started up on the same day, and run continuously. Then, a year later, if there were no heat lost to the environment, the lake would be 0.6°F warmer, more than a degree warmer the next year and so forth. But of course this picture is not realistic. Some of the heat added during the first year will be lost to the atmosphere through evaporation, long wavelength infrared radiation and by direct transfer of heat. The question is, how much? A simplified **heat budget** for the lake is shown in Figure 8–1. The net atmospheric and solar radiation into the lake and the heat added from the power plants are not dependent on the lake temperature. Other terms, such as the rate of heat transfer by evaporation and long wavelength radiation and the direct heat transfer, *are* dependent on lake temperature and increase when the lake temperature increases, as shown in Figure 8–2. The lake temperature

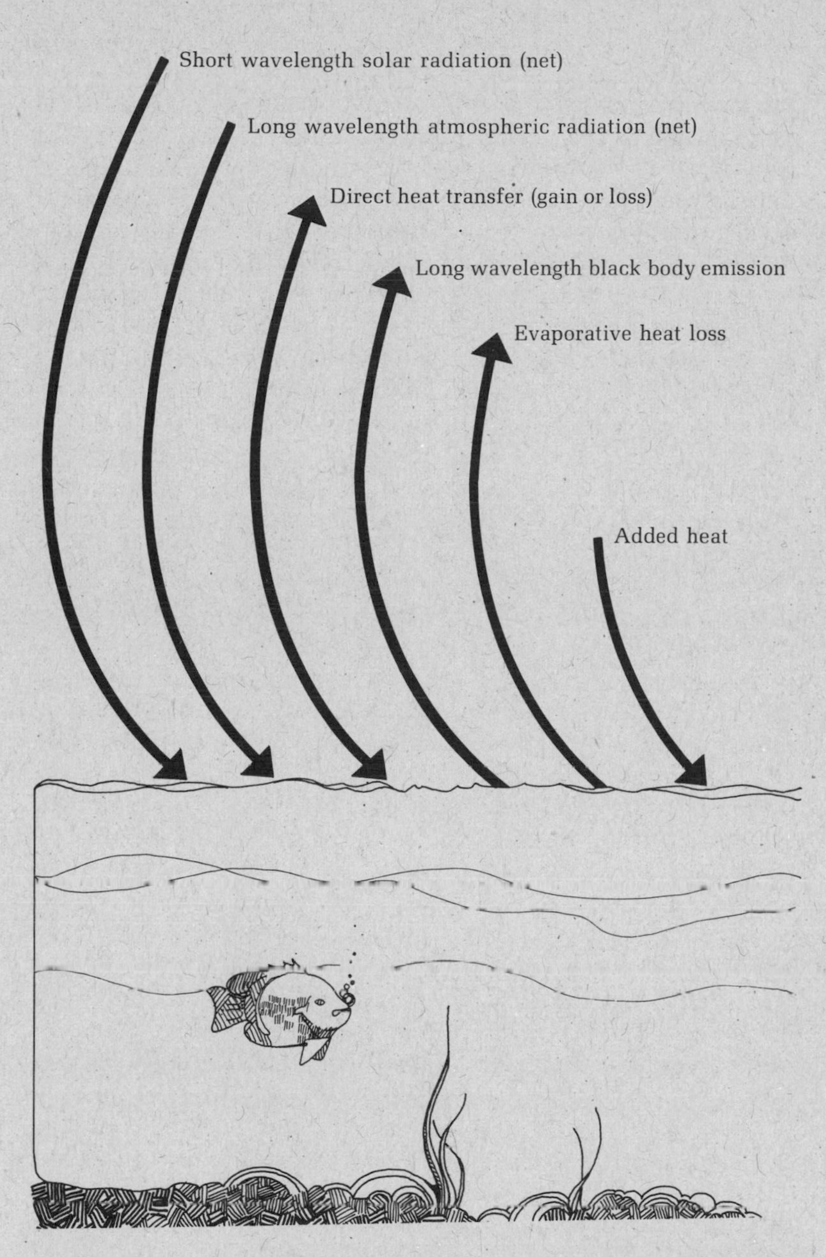

FIGURE 8–1. *Simplified heat budget for a water body. The net solar and atmospheric radiation absorbed is corrected for the small portion which is reflected. Direct heat transfer might go in either direction, depending on the relative temperatures of the air and water.*

will remain constant as long as the heat coming in balances the heat going out. As the seasons change the lake temperature must change because the terms in the heat budget change. For example, in the summer there is more solar and atmospheric radiation going in, and direct heat transfer is *from* the atmosphere *to* the water. The lake therefore warms up until the rate of evaporative loss and long wavelength radiation outward compensate the heat transfer in, and a balance is again struck. Now, if the heat from the power plants is added to the balanced system, how much will the lake have to warm up so that the heat loss rate again balances the heat input rate?

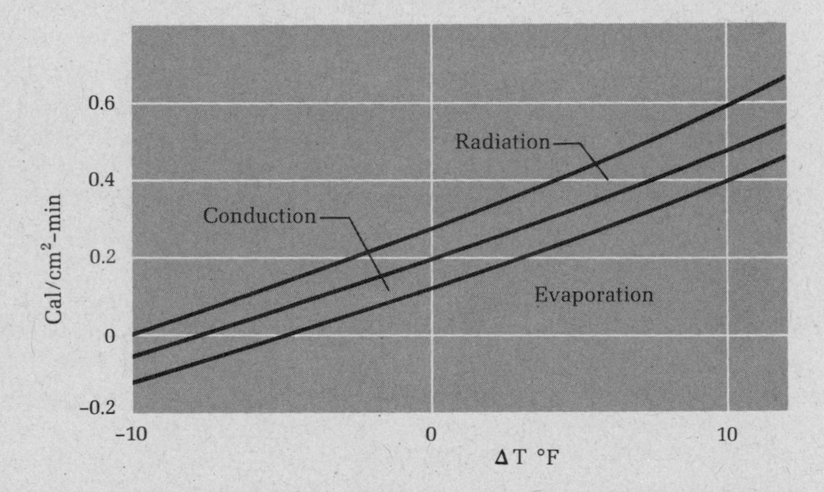

FIGURE 8-2. *Variation in heat transfer from a lake surface with a change in lake temperature. ΔT represents a change in lake temperature as compared with some standard condition chosen as reference. The figure is meant to illustrate how the heat transfer varies with a change in lake temperature. Most of the variation comes from the change in evaporation rate with temperature.*

This question cannot be answered exactly, because the detailed heat budget for a body of water as large as Lake Michigan is not available. However, studies have been carried out on other lakes, and from these it is possible to make some very good estimates. For simplicity, let us assume that the heat added to the lake by the power plants is uniformly distributed over the area under consideration. Over a period of time it will be so distributed, even though the additions are being made at specific, localized sites. This uniform distribution would produce a heating rate of 0.00092 calories per square centimeter of surface per minute. The rate at which heat is transferred out of the water, by evaporation, radiation and direct transfer, is very much larger than this, about 0.2 calories per square centimeter per minute. More importantly, the rate of heat transfer increases rapidly with temperatures as illustrated in Figure 8–2. Based on the data available, it can be concluded that an increase of only about 0.05°F in the lake temperature would be sufficient to counterbalance the effect of added heat from the power plants.[2]

This particular estimate is not dependent upon whether we assume that the waters of the epilimnion and hypolimnion become thoroughly mixed. Of course, in the cold part of the year mixing does occur. The estimate does indicate that the long-term increase in average temperature of the lake should be no greater than about 0.05°F—about 0.1°F if we double this to err on the side of safety. This seems like a rather small temperature increase, and indeed it is. Because mixing is not likely to be very complete until a fairly long time has elapsed, some sections of the lake may attain higher average temperatures with consequent increase in biological activity. Proper design of cooling water discharges could, however, circumvent potential difficulties of this sort. It must also be recalled that the estimate applies for a particular quantity of added heat. If the added power capacity on the lake shores were doubled or quadrupled over the assumed 10,000 megawatts, the estimated temperature changes would be up to 0.2 and 0.4°F, respectively.

Even a small temperature increase produces some increased biological activity. As a rough estimate, a 1°F rise in temperature corresponds to about 5 percent increase in the rates of biological processes. Therefore, even the small increase estimated here carries a penalty in the form of some increase in the rate at which the lake eutrophies. This penalty must be balanced against the social benefits, if any, which accrue from the increased power capacity. The costs of not having the power might be much higher than the costs of eliminating or drastically reducing the level of nutrients now being added to the lake. There is little doubt that eutrophication would be retarded much more by a vigorous cleanup of effluents and runoff into the lake than it is accelerated by a temperature increase on the order of a tenth of a degree.

It is interesting to note that if the total heat rejection of the new electric generation plants were dissipated through the use of evaporative cooling towers, a total of 68 billion gallons of lake water per year of consumptive use would be involved. This is about 16 percent of the entire consumptive use of the metropolitan Chicago area. There has been concern for a number of years that the consumptive use of water from Lake Michigan might grow to be excessive and lead to disastrously low lake levels in periods when the runoff has been low for a number of succeeding years. Consumptive use at this level would add to what is already a serious problem. Even if cooling towers are not used, however, about half the heat added to the lake from the power plants will be dissipated through evaporative cooling, so there will be a substantial consumptive loss in any case.

Now let us consider a rather different situation. Cayuga Lake, one of the Finger Lakes of New York State, is 38 miles long, has a mean width of 1.7 miles and a total surface area of 66 square miles. Despite this rather small surface area its maximum depth is 435 feet and its mean depth is 179 feet. Cayuga Lake represents a classic case of a highly stratified lake during the summer months. The waters of the lake do not have a rapid turnover rate; the mean flushing time is about 9 to 12 years. The lake is relatively infertile, which means that the quantity of algae and other plankton organisms is not excessive. Observations over the past 40 years indicate, however, that eutrophication is proceeding at a measurable rate. Cayuga Lake and the other Finger Lakes are splendid recreational and esthetic assets. Their premature decay through accelerated eutrophication and pollution would be a great tragedy.

In about June, 1967 the New York State Electric and Gas Corporation announced its intention to construct and operate an 830-megawatt capacity nuclear-fueled power station, to be situated adjacent to the site of a smaller existing fossil-fueled plant, on the shore of Cayuga Lake.[5,6] In the company plan, water would be removed from the hypolimnion, at a depth of about 100 feet, warmed some 20 to 25°F, and returned to the epilimnion. Temperatures in the epilimnion range from 50 to 73°F during the May to November period of stratification.

Thus the water would be discharged to the upper layer of the lake at a temperature not too much different than the surface waters. Standards announced by the New York State Water Resources Commission in August, 1969, require that the cooling water discharges be made near the surface, and that they not raise the surface temperature more than three degrees beyond a radius of 300 feet from point of discharge. The cooling system proposed by the company could presumably meet these requirements without difficulty.

Active opposition to the proposed new siting of the power plant began to form within six months of the initial announcement. By May of 1968 a group of Cornell University faculty, mostly from the biological sciences, had prepared a document detailing the nature of the relationship between the lake and the proposed new plant, spelling out some of the known and possible effects, and suggesting alternative cooling systems. The activities of this and other citizen groups led state legislators to introduce bills providing for safeguards against thermal and radionuclide pollution of lakes. Two of these bills, passed by both arms of the state legislature, were vetoed by the Governor.[6] In the view of the Citizens Committee to Save Cayuga Lake the state standards announced in 1969 are too lenient.

In response to the criticism of its initial proposal, New York State Electric and Gas Corporation announced in April, 1969, an indefinite postponement of the company plans to build the nuclear plant. The purpose of the delay was to permit additional research on cooling systems and consideration of the economic effects of such systems. Cornell's Water Resources and Marine Sciences Center released in November, 1969, an extensive analysis of Cayuga Lake, and the possible effects of the proposed power station. The report concluded that the proposed plant initially would have no acute effects on fish population, but did find that it would probably stimulate the growth of plant organisms.[7]

To date there is no indication that the company will significantly alter its original plans for the cooling system. The local Citizens' Committee continues to oppose approval of the company plans by the state commission. The utility has declared a moratorium on plant construction pending some indication of whether the Federal Water Quality Administration will approve the proposed state standards. Completion of the nuclear plant is not now expected before the late 1970s.

This particular case involves some very interesting and

knotty problems. The company has been strongly opposed to the use of cooling towers, ostensibly on the grounds that experience with cooling towers in a climate so far north and with the particular climatic conditions which prevail in the Finger Lakes region is lacking. The consequences for the lake of the cooling system proposed by the company have already been outlined: increase in the volume of the epilimnion, a longer growing season with a higher rate of biological production, increased oxygen demand on a smaller hypolimnion. The cooling water would be discharged at a rate of 500,000 gallons per minute. This amounts in a year's time to one tenth the total volume of the lake. Taking into account the seasonal variation in stratification, it would increase the volume of the epilimnion in October by about 20 percent.

There is another way of viewing the situation which indicates the magnitudes of the effects which may be expected. Suppose we go through the exercises carried out in discussing Lake Michigan. First of all, if all the 1600 megawatts of waste heat generation from the new plant is put into the lake, assuming there is no increase in heat transfer to the environment, the heat added would raise the temperature of the lake about 2.5°F in one year! This assumes that the heat is distributed evenly throughout the lake. If it were assumed that the heat were confined to the epilimnion, a much higher calculated temperature rise would result. But of course the lake does not retain all the added heat. The lake temperature must go up; as it does, however, the rate of heat transfer to the environment increases, and a new equilibrium surface temperature is attained. Again, the question is, how much higher is the new temperature? A quick comparison of the numbers shows us that the calculated increase in temperature will be much higher than in the Lake Michigan example. The one power plant on Cayuga Lake has a heat rejection rate about 1/12 of the total assumed for the Lake Michigan installations, but the lake area is only 66 square miles, as opposed to the assumed 12,000 square miles for the latter. If we make the same assumptions about the rate of heat transfer from the surface to the atmosphere as in the Lake Michigan calculation,[8] the new equilibrium surface temperature when the plant is operating must average about 0.7°F higher than at present—1.4°F if we double this to err on the side of safety. In other words, the thermal effect on Cayuga Lake of the one new plant will be more than ten times greater than the effect on Lake Michigan of ten larger plants! Although it cannot be claimed that this estimate is very precise, it is based on quite reasonable premises. Taking into account that the annual temperature variation in the surface waters of Cayuga Lake is not unlike that in Lake Michigan, and that the air in the Finger Lakes region is fairly cool and moist, the estimated rise in surface temperature is, if anything, probably on the conservative side.

The increase in surface temperature resulting from the added power plant heat will not be constant throughout the year. One expects that the increase might be smallest in late summer and early fall when evaporative rates are relatively higher. The seasonal variation will be affected by the details of thermal mixing, the time required for distribution of the cooling water in the epilimnion, and other factors. The major point, however, is that if the surface temperature must be increased an average of 1.4°F to provide the compensatory increase in rate of heat loss, the body of the epilimnion during periods of stratification and of the entire lake during the mix ing period must be increased at least this much and possibly more. Thus it seems quite probable that the proposed power plant would have the effect of very significantly increasing the temperature of Cayuga Lake and thus contributing to an increased rate of eutrophication. It should be noted that this would occur regardless of the manner in which the company withdrew and returned water to the lake. An enhanced rate of evaporative cooling resulting from a relatively hot mixing zone (forbidden by state standards) would help to reduce the load on the rest of the lake. The benefits of this strategy would, however, be marginal and might result in considerable disruption of the lake's ecology. It therefore appears that cooling towers of some form are required. Use of cooling towers during the months in which the lake is stratified would be of considerable benefit in avoiding the possible adverse effects already alluded to. During these months the likelihood of fog and icing resulting from use of evaporative cooling would be minimal. However, it is important to recognize that even if the lake receives the full thermal load only during the coldest six months of the year, the heat which it thus accepts will affect the average surface temperature significantly during the warm season. Use of a combination of evaporative and dry cooling towers over the entire year would seem to be the only really safe course and even then it must be ascertained that the lake could bear the burden of a substantial consumptive loss from use of evaporative towers. In this connection, the company plan would also involve a very substantial evaporative loss, since about half the heat added to the lake would be dissipated at the surface through evaporation. In a year's time the lake would lose from 4 to 5 billion gallons. This corresponds to a 4-inch layer of water over the entire lake surface. If the New York State Electric and Gas Corporation were made to pay for the water thus *consumed* (not just used), the costs of even a dry cooling tower might seem less prohibitive.

Our third and final example involves a power plant siting on a river. The Vermont Yankee Nuclear Power Corporation was formed as a Vermont corporation in August, 1966. It is jointly owned by ten investor-owned New England utility companies; the major holders are Central Vermont Public Service, Green Mountain Power Company and New England Power Company, in that order. A few days after its formation the company announced that a 540-megawatt capacity nuclear-powered steam electric plant would be constructed near Vernon, Vermont, on the Connecticut River. The site is a few miles north of the Vermont-Massachusetts state line.[9]

In accordance with the provisions of the Atomic Energy Act of 1954, application was filed in December, 1966, with the Atomic Energy Commission (AEC) for permission to construct the plant. The Division of Reactor Licensing of the commission issued its safety evaluation in July, 1967. Following this the commission scheduled a prehearing conference for later in July, with formal hearing in August, 1967.

The states of Vermont and New Hampshire both had statutes in effect at that time providing for licensing of activities which result in discharge into intrastate waters. The Vermont Department of Water Resources in December, 1966, requested the company to make application for such license. This was followed by a demand in July, 1967 that it do so. The states involved requested and obtained a continuance of the AEC hearing until September, 1967, in order to prepare material and arguments relating to the thermal effect of the proposed plant on the Connecticut River.

The plant design as originally submitted to the AEC apparently provided for a single-pass, direct-discharge cooling system, utilizing the waters of the Connecticut River. The design specifications were such that the condenser could be expected to require about 780 cubic feet per second of cooling water; the water would be heated 20°F in passage. A cooperative program between the Vermont Department of Water Resources and the U.S. Geological Survey had determined that the minimum monthly flow during the 1945–1965 period was about 1030 cubic feet per second. The minimum average flow for a single day, however, was as low as 100 cubic feet per second, and three-day averages often were on the order of 200 cubic feet per second and lower. The Connecticut River is dammed at points above and below the plant site for production of hydroelectric power. In addition the river must serve as a navigable stream, assimilate municipal and industrial wastes and serve recreational demands. The highly variable flow levels at Vernon resulted from attempts to satisfy all these needs, but were primarily the result of hydroelectric power generation. Accordingly, Vermont Yankee obtained a guarantee from the New England Power Company, one of its member companies and operator of the hydroelectric power facility, of a minimum flow of 1,200 cubic feet per second at Vernon.

The record of this particular case indicates quite clearly that the Atomic Energy Commission was unwilling to consider thermal pollution effects resulting from the plant as a criterion in its licensing evaluation. At the same time, it might have been obvious to the responsible officials of Vermont Yankee that they were not going to get by with a simple, single-pass flow system. In addition to the request (and then demand) from the state of Vermont for a filing for license from the state, the company had the benefit of an opinion of the Bureau of Fisheries of the Department of the Interior, as part of the reactor licensing report, that the thermal effects of the proposed system would be detrimental. A 20-degree heating of two-thirds of the stream flow would, especially in summer, quite thoroughly clear the river of many species of fish and would completely nullify attempts underway to reestablish the river as a ground for American shad and Atlantic salmon. At a hearing before the Vermont Water Resources Board in August, 1967, the company, as represented by its chairman, Mr. A. A. Cree, indicated a willingness to meet any reasonable water temperature and temperature rise standards which might be imposed.

It seems evident from the alacrity with which the company reponded to formal proceedings on the part of the Vermont state agencies that it anticipated action at the state level, and was prepared to meet demands for provisions to reduce thermal loading of the river. It also seems evident that, in the absence of these demands, the plant would have been constructed as originally planned.

In December of 1967 the company presented a plan for installation of a flexible system of two mechanical draft, evaporative cooling towers which may be operated as either an open or closed system. That is, it may recirculate the cooling water in whole or in part, or may return it to the river after a single passage through the cooling towers. The system as designed should permit the company to operate within the temperature and temperature rise limits as set by the states of Vermont and New Hampshire. These limits permit the company to discharge heated water which could cause a 9° change in river temperature in winter, whereas in summer essentially no increase in river temperature is allowed. It has been estimated that the cooling tower system, which is required to be in operation year round, will add about $6.5 million to the total plant cost of about $130 million. Vermont Yankee is scheduled for completion in mid-1971.

The Connecticut River is not the unspoiled home of trout and spawning ground of salmon that it once was. The intrusion of still another of man's enterprises into the river ecology does not seem of such great moment when the river is already so much used. For the moment everyone seems satisfied that a reasonable compromise has been achieved, although there are worries about the possibly heavy incidence of fog and icing during winter.[10]

It is anticipated that about 500 new plants of at least 500-megawatt capacity will be put into operation throughout the United States by 1990. Of these, up to about 100 will be large nuclear plants with 1000–4000-megawatt capacity. It is evident from the examples we have considered that it will not be easy to find suitable locations for all these additional installations. It might appear at first that the problems of heat disposal could be taken care of by pumping the heat to some remote location. There is, however, no well-developed technology for transporting heat over great distances. It must, therefore, be disposed of at the site of power generation. The alternative strategy would seem to be to locate the power stations where there is adequate space for building artificial lakes or cooling ponds, or where natural water supply of sufficient capacity is available. Setting aside the difficulties created by the increasing scarcity of land and the limited availability of natural waters, remote siting of power stations creates special problems in its own right.

Transmission of electricity over considerable distance is a necessity regardless of plant location, because of the need for transfer of power between regions to meet power shortages and emergencies arising from equipment failures, storm damage and unexpected peak demands. Transmission of power from one area to another also helps to reduce the total generating capacity needed and thus achieves economies in capital and operating costs. The technology of power transmission is quite complex. Use of very high voltages has made it possible to transfer increasingly larger amounts of power.[11] At the same time the costs of transmission have increased as the systems have become more complex and as the land required for the towers has increased in value. Furthermore, the transmission systems frequently mar the beauty of the countryside; in some instances they may have intruded on land set aside for recreational use, or as a wildlife preserve. It seems certain that opposition to particular locations for transmission lines will be increasingly forceful in the future.

Underground location of transmission lines, which might circumvent at least the esthetic disadvantages of the above-ground installations, is not generally feasible with the present state of the technology. It is not possible to employ voltages as high as those now commonly used in the above-ground systems. Even for the highest voltages which can be employed, the underground systems are from 10 to 40 times more costly.[12] One of the more interesting possibilities for underground transmission is the use of a cryogenically cooled cable (page 67). However, the technology of a cryogenic cable system is not well developed and the prospects for an operational system of this kind in the next few decades are not bright.

Remote siting is frequently put forward with enthusiasm by power company executives concerned with providing the capacity to meet power demands. While it may serve in the short run to ameliorate certain critical conditions, it by no means provides a satisfactory long-range answer to the problems created by the apparently insatiable appetite of the economy for more power. It would be unfortunate if a planning policy based on remote siting or any other alleged panacea for the environmental ills created by power demand were to divert attention from what should be a preeminent consideration: there is a limit on the growth of national energy consumption. The rate at which any form of energy consumption can expand should be determined by considerations of prudent management of our environmental and human resources. In these terms the growth rate of the electric power industry is too high. Its eventual size is related to the total allowable rate of consumption of energy in whatever form. Long before that stage is reached, however, there must be very substantial breakthroughs in the techniques of heat management.

At first sight it would seem that all of the furor over waste heat is quite unnecessary. If the heat represents energy, why not put it to good use? One can think of many possible uses of heat. It is a fact, for example, that the waste heat from electric power plants represents sufficient energy to heat all the private homes in the United States. Before we consider this and some other possible uses of the waste heat, however, we must reflect upon the nature of the product we have to work with.

It was pointed out earlier that the efficiency of a steam electric plant is improved by maintaining the condenser at as low a temperature as possible. Therefore, the electric utility must try to pump water through the condenser rapidly enough so that it does not increase in temperature any more than necessary. The result is that a great deal of water gets pumped through and is heated from 10 to 30°F as opposed to allowing a smaller quantity of water to heat up much more. The total **amount** of heat transferred is the same, or nearly so, in the two cases.

In order to optimize the useful work obtainable from heat it is necessary to make the temperature difference between one part of a system and another as large as possible. The unfortunate fact about the heat stored in the cooling water is that it does not leave the cooling water very much hotter than before; surely, it is enough hotter to cause serious mischief, as we saw in the last chapter, but not enough to provide the basis for extracting useful work. This, indeed, is why we refer to it as "waste heat." But if it cannot be used directly to perform work, such as the running of an engine, etc., perhaps there are other ways in which it might be used.[13]

By allowing the cooling water to get just a little hotter than it does in systems as they are presently run, it might be possible to utilize the heated water for home heating, or for keeping sidewalks and roads free from ice in the winter. This might be feasible in a carefully integrated plan in which a new city is being built along with the power plants which will serve it. Such utilization of cooling water discharge at present, however, in cities which are already on the point of collapse because of the welter of underlying non-coordinated services and municipal functions seems quite impracticable. There is the further problem that the need for the waste heat is at a minimum in the hot months, just when it is most critical that some use be made of it. The development of improved designs of heat engines might make it possible to employ the heat in the hot months to operate air conditioning systems, but there is little prospect of any immediate utilization of this kind.

Desalination of seawater by evaporative techniques requires some heating of the seawater. Although cooling water discharge is not hot enough in an ordinary installation to be very effective, some compromise between maximum efficiency in cooling and effectiveness in desalination might provide both electrical power and fresh water at reasonable prices, especially in desert areas. Some limited use is being made now of nuclear reactors for desalination, but large-scale development is not imminent.

It has been suggested that warmed waters discharged from power plants might be used to increase biological productivity in a variety of contexts.[13,14] For example, lobster might grow more rapidly in warmer waters than they encounter in their breeding grounds off the Maine Coast. Similarly, oysters might be cultured in areas bathed in power plant discharges. The yields of various fish such as pompano and flounder might be improved by cultivating them in warmed waters. Preliminary studies are underway to determine the feasibility of putting warm water to use in this way. For the most part, this use of power plant cooling waters is restricted to locations on the sea coasts.

The efficiency of waste treatment varies considerably during the course of a year in areas where the temperature varies over a wide range. It would be possible to improve the overall performance of a municipal sewage disposal plant by utilizing cooling water discharge to preheat raw sewage, and to maintain settling tanks at a constant warm temperature. Such a use would not involve any mixing of the warm discharge water with the sewage; it would simply flow around the vessels and pipes in the sewage treatment plant. This use of discharge waters would be difficult to implement in established installations, but it might be incorporated as part of a new facility.

In areas which require large quantities of water for irrigation and which are not excessively hot, it might prove beneficial to apply heated waters from power plants directly to the land, using any one of several commonly employed methods of application. Use of heated waters might prevent frost damage at some times of the year and might generally stimulate growth. Experiments with use of discharge waters are underway in the states of Oregon and Washington.[15] If they are successful, this application might be developed sufficiently to absorb a considerable fraction of the total discharge from new steam electric plants in the Pacific Northwest for a number of years. As yet, however, there are no results available.

In addition to all the possible uses just discussed, many others have been suggested. One of the more grandiose is a suggestion that the Saint Lawrence Seaway could be maintained in ice-free condition by judicious location of electric power stations along the waterway. The amount of thermal discharge required, however, might not be in accord with good conservation practice during the warm part of the year. Furthermore, it would be difficult to provide assurances that the ecology of the seaway and associated waters would not be adversely affected.

The search for beneficial uses for the heat values in power plant discharge waters is worthwhile and it should go on. At the same time, it must be admitted that there are no large-scale uses in sight. The most promising at the moment is use in agriculture, but it is too early to tell whether application of heated waters in irrigation might not produce undesirable results not forseen at this time. Even assuming the best, the application of the technique is limited. Similarly, use of warmed waters in fish and aquatic plant cultivation might take on increased importance if favorable results are obtained in pilot studies. Some scientists suspect, however, that while fish flourish for a part of their life span in warmed waters, their reproductive behaviors may not be improved and may even be impaired by exposure to higher temperature.

We must conclude, therefore, that while substantial benefits may someday be forthcoming from uses of the heat in discharged cooling waters, it is very unlikely that a major fraction of the waste heat which will be generated in an ever-increasing quantity will find gainful employment. Thus we cannot look in this direction for much relief from the pressures of thermal pollution. From the standpoint of our concerns about the total rate of energy generation on a global and possibly even a regional basis, the effect on climate is more or less independent of whether some "use" is made of the heat generated before it is unloaded in one form or another on the atmosphere.

9 a new conservatism

The preceding chapters have dealt with the earth's radiation balance and with a number of factors important in affecting climate on a planetary, regional and even local level. In addition, the effects of heat generation on other components of the environment, notably natural waters, have come in for some discussion. I hope that I have been able to convince the reader that it is often possible to make fairly good estimates of the effects on climate of man's activities. But these activities are diverse and their effects are diverse. Some of the possible consequences of our inadvertent tinkering with the planetary climate seem less threatening than others; some are of more immediate concern than others. We need at this point to synthesize what has gone before and to fit our results into the larger perspective of the contemporary views of climatologists.

We have seen that the earth's radiation budget is in balance, or very nearly so. Solar radiation which enters the atmosphere is either scattered or reflected back into space, or is absorbed by the atmosphere or at earth's surface. The energy input is balanced by the long wavelength radiation outward into space. The climatic conditions at the surface are determined primarily by the temperature profile in the atmosphere, extending upward from the surface. This temperature profile is affected by a number of variables, including the extent and nature of cloud cover; the concentrations of infrared-absorbing gases, particularly water and carbon dioxide; the quantity and distribution of particulate matter; and man-made heat added to the atmosphere at the surface.

To summarize very roughly, we may say that the predominant effect of human activity seems to be in the direction of warming the planet. Increasing carbon dioxide content and a rapidly increasing level of energy generation at the surface both exert a warming effect. The projected time scales for the growing influence of these two major contributors to warming are rather similar: unless there are significant departures from the projected usages, the period 2050 to 2100 should bring the first unmistakable signs of a major man-made warming trend. On a shorter time scale the climatic significance of the SST flights is rather conjectural at this point. If it should turn out that the planes produce a substantial increase in high level ice crystal clouds, they might contribute to a warming. On the other hand, they may add to the atmospheric load of particulates and in this way may exert a small cooling influence.

To counter the warming effects just mentioned there is only increased atmospheric turbidity. Actually, the real climatic influence of particulate matter is somewhat more conjectural than, say, the influence of increased carbon dioxide.[1] It cannot be said at this point that mankind could not raise enough dust to counter the heating effects of carbon dioxide and energy generation, but we may ask whether we wish to leave this possible necessity as our heritage to future generations.

It is not necessary that we have highly quantitative estimates of the long-term effects of all these man-made influences before deciding whether corrective measures are called for. The analyses presented here, based on all the available information, tell us that we could be headed for trouble. Any one of a number of man-made influences is of itself capable of seriously affecting the planetary climate. To argue that we should not worry about any of them because they are liable to balance out in the long run would be absurd even if they did in fact turn out to be balancing to a nice degree. Who would advocate rendering our atmosphere more turbid because that would permit us to continue heating up the environment?

It is equally absurd to argue that we should not concern ourselves about man-made influences since other factors have obviously operated in the past to alter the global climate.

There is certainly very strong evidence that the earth's climate has undergone great changes during the past millions of years.[2] The climatic changes which have occurred during the past 150,000 years or so,[3] hypothetically ascribed to changes in solar activity, perturbation of the earth's orbit and dozens of other possible influences, have been of such a magnitude that the man-made influences projected for the future could easily become over-riding. For example, the average terrestrial temperature seems to vary up and down with a period of about 80 years. The variation is thought to be due to fluctuations in some component of solar radiation; probably not the portion in the visible and near-infrared range which is measured by the solar constant, but in the high energy ultraviolet ranges. In any case, the indications from recent work are that the earth's temperature will reach a minimum in about 20 or 30 years as a result of this factor and then will rise to a new maximum in about 2050 or so. The magnitude of the temperature variation from this effect is not clear, but it is probably not more than 1 or 2°F. Man-made influences are simply superimposed on these naturally occurring variations in temperature. It could easily happen that a man-made warming influence, coming on top of a naturally occurring upward swing in temperature, could trigger a drastic change in global climate.

Climatologists are not very certain about what forces are important in causing climatic change. The climatic history of the earth is known only very sketchily and only for the most recent one-tenth of the earth's history. It is not even certain how many glacial periods made up the last ice age. At present we seem to have a climate which is intermediate between the cold glacial and warm interglacial extremes, but it is not clear toward which extreme we are swinging. Really useful scientific records such as weather and atmospheric data have been regularly kept for only the past 50 to 100 years. It is now realized that many of the large-scale movements of air in the atmosphere are unstable and subject to change. These changes may occur over a long time, perhaps a hundred years or more. Similarly, it is possible that the sun's radiation varies in ways not yet obvious to us, in addition to the well-known 11-year variations in sunspot activity. The relatively short time over which modern, high quality scientific data have been accumulated does not yet, therefore, permit climatologists to see very clearly how these changes may occur, or what they lead to.

The inherent instability of the atmospheric circulations suggests that climatic changes, when they do occur, may be rather sudden. There is evidence that in the past the earth has undergone many catastrophic climatic shifts and we have no assurances that it will not again. But as man learns more about the radiation budget and the details of all the many interactions between land, sea and air, it should become increasingly possible to understand and anticipate climatic change. In the absence of that kind of understanding, we are not clear on how important human activities might become in forcing climatic changes. The estimates put forth in this book, for example, are based on the assumptions that all the factors which determine the climate remain the same and that our man-made inputs, because they are relatively small additions to the system, do not disturb these other factors. But it is very likely that climatic changes will bring about changes in the other factors which of themselves could produce much larger alterations in environmental conditions than we calculate for our initial influences. In short, the relatively small influences we anticipate in some cases might be amplified far beyond our expectations, plunging the planet into a deeply interglacial period, with complete melting of all the ice and a 150- to 200-foot rise in the water level of the oceans, or into a frigid ice age.

It is just possible that if man continues along the present course, expanding his energy output in accord with the projections, using up fossil fuels on the projected time schedule, and steadily increasing the atmospheric turbidity, no major climatic changes will occur. There may be a small net warming or cooling, but nothing drastic. It may be a bit cloudier on the average than now; the skies will be hazier and less blue than now. There may be serious problems in deterioration of the environment as a result of waste heat dissipation; that problem is distinct from climatic change and will be discussed separately. Insofar as world climate is concerned, things just might be all right. The evidence suggests, however, that such a (relatively) happy outcome is very unlikely.

Using the rather conservative estimate made in Chapter 4, the increased carbon dioxide level will be responsible for a rise of about 4°F in the average temperature by about 2050. Added to this there will be an increasing contribution to higher temperatures from man-made energy dissipation. Atmospheric turbidity will doubtlessly intensify beyond the present level, but closer attention to pollution control will in all likelihood limit the increase to such an extent that loss of insolation due to turbidity will not effectively counter the warming influences. No one can say with certainty whether a man-made alteration in surface temperature on the order of 4°F or more, serious enough in itself, might trigger a much more drastic change in the climatic regime.

These prospects present mankind with a new kind of challenge, appearing for the first time in man's history in the twentieth century. In addition to the need for avoiding nuclear war and for limiting population growth, energy control will become a necessity if civilized society is to be maintained for an extended time. The problems of finding enough food and other resources to sustain the race are strongly coupled to the population problem. The per capita food consumption is limited to some range of intake which represents good nutrition. The use of mineral and other resources can be related to population level given a change in economic philosophy to provide inducements to recycling. But whereas all these other factors related to maintenance of a healthy, technologically advanced civilization can be related to population level, energy consumption cannot. There is no evidence as yet of economic or technological forces which might limit per capita energy consumption. It seems possible to conceive of a Buck Rogers-like future, in which man lives in completely enclosed cities on a planet grown alien, insulated against the outside environment by the continuous application of man-made energy, and sustained within by heavy per capita energy investment. One can only speculate on how many people such a civilization could support. The population density could be quite high. All food would be synthetic; perhaps even the sun's rays would be replaced by artificial light sources, or by alternative chemical synthesis of foods.

While we may well leave to the future the design of its own world and not attempt to anticipate or judge it, we cannot escape the fact that decisions made about the future course of world society during the century ahead, and beginning now, will determine whether a society survives to build a life style of *any* description. We may succeed in making large portions of the earth uninhabitable before man is technologically prepared to cope with the consequences. Thus energy conservation belongs along with population control at the forefront of problems of concern to the entire world society.

Kenneth Boulding points out in a very perceptive and thought-provoking article the difficulties inherent in arousing sustained current interest in the future and in the prospects of some ill-defined posterity.[4] Nevertheless, the nerve and fiber of a society are linked with its sense of place in the flow of human history. This perennial need for identification with the past and sense of connection with the future has placed especially great stress on contemporary society. Regardless of what may be said to the effect that our time is no different than other periods in man's history which have produced radical change, we recognize that we are different in the very important respect that we have the capacity for totally destroying human civilization. Even the most ardent slaughter of earlier times could not effect the degree of destruction of established social order, to say nothing of purely physical destruction, that is now possible merely by inadvertence, carelessness or an isolated instance of madness. This knowledge of man's power of self-destruction produces at once a heightened sense of isolation from the past, and renders the future more tentative, more ephemeral. In the past man knew that there would be a future; now he cannot be so sure.

The present period of human history is unique also in that mankind is reaching the limits of growth on the planet. While the pace of technological change continues to increase, the physical limits in terms of population and energy consumption are clearly in sight. If we make a graph of some important index of technological advancement, such as per capita energy consumption, against time, we get something like that shown in Figure 9-1. A graph of human population vs. time would also have a similar S shape. The present lies in the steep section of the curve. We live in a unique period of human history, therefore, in the sense that the rate at which change is occurring is greater now than it has ever been, or can ever be in the more distant future.

Individually we have a tendency to discount the future. We do this even in our own lives, as illustrated so cogently in Stephen Vincent Benét's story, "The Devil and Daniel Webster." The future as it affects posterity is discounted even more. Discounting of the future operates in much the same manner as collecting interest on money lent. Something which will occur next year is of less concern, say about 5 percent less, than the same event occurring today. At this rate of discounting, a time fourteen years into the future weighs only half as heavily in our concerns as the present, and the year 2050 is of not much concern at all. But in a sense, our terrestrial immortality is vested in our progeny and institutions. To the extent that we identify with them we have concern for the future beyond ourselves.

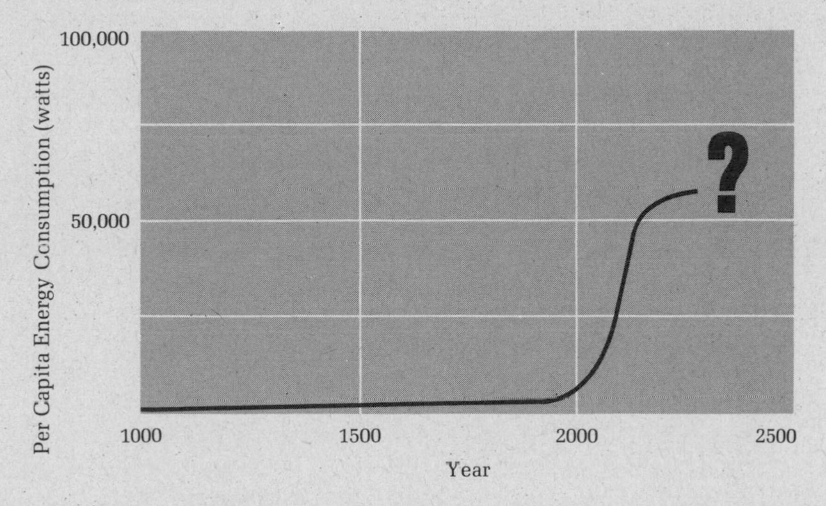

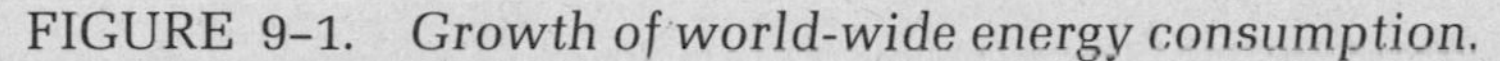
FIGURE 9–1. *Growth of world-wide energy consumption.*

To cope with the problems which beset us and to plan effectively for the future, we need first of all a clear perception of our unique place in the continuum of human history. Secondly, we must generate from that perception a feeling of connection with the future, a sense of responsibility for what we leave to posterity. But humanity may simply not be able to rise to the occasion. If the many easy pathways to disaster are to be avoided, mankind must divest itself of much outmoded conventional wisdom. John Platt has written about how we must organize to begin dealing with the problems which must be solved.[5] A radical redirection of the technological capabilities and creative energies of the society is necessary to marshall sufficient capacity for solving the most urgent problems before they overwhelm us. Unfortunately, it is not likely that there will soon be a sufficiently widespread recognition of the urgency of many of these problems, or agreement on the actions required to begin working on solutions. When one considers the intellectual and moral qualities of the political leaders of the major nations, the hopes for dynamic and enlightened initiatives toward international cooperations and accommodations are dim. Nevertheless, there are promising signs here and there. A few feeble movements toward meaningful international accords on such important items as arms limitation, non-proliferation of nuclear weaponry, the halting of atmospheric testing of nuclear weapons, all give some reason for hope. It may be that widespread acceptance of controls on atmospheric pollution, fossil fuel consumption and energy consumption, less "loaded" topics than some others, might help to set the stage, establishing precedents of value in dealing with the hotter and more immediately urgent problems such as population control and arms limitations.

In a report to the forty-seventh session of the United Nations General Assembly, in 1969, the Secretary-General discussed the problems of the human environment. These were categorized as problems of human settlements, territorial problems, and global problems. A United Nations conference has been called for 1972, in Sweden, for comprehensive consideration of these problems. It is of interest that the problems identified in the Secretary-General's report as global in character include increasing carbon dioxide and particulate matter and the discharges of heated waters into estuaries of coastal waters on which the productivity of the oceans is dependent. No mention is made, however, of the problems associated with increased overall heat production. It is very much to be hoped that this topic will receive intensive consideration by the Commissions on Environmental Pollution and on Environmental Aspects of Economic and Social Planning.

Let us consider what sorts of international controls and accommodations would be required to limit the world-wide climatic influences of atmospheric turbidity, carbon dioxide and heat rejection. Atmospheric turbidity is the easiest one of these to deal with. The sources of particulate matter derived from man's activities can be determined and remedial measures applied. Industrial processes probably account for a large fraction of the increase observed during the past 50 to 60 years. Control of particulate matter in exhaust gases is now being effected in the United States in response to public clamor for cleaner air. During the past few years there have been immense improvements in techniques for controlling particulate matter, and we can expect continued advancement in this area. These techniques will be transferable to other nations. Control devices which are made integral with new installations as they are constructed in developing nations should cost less and be more widely acceptable than measures which must be tacked on to existent offending sources. The motivation for controlling particulate matter derives from the simple fact that dirty air is not good for people and they know it. The massive cleanup of air now underway in the United States does not stem from an awareness that this nation has been adding to the turbidity of the atmosphere and thus possibly influencing world climate. The fraction of particulate matter from industrial stacks which remains in the upper atmosphere for an extended time is miniscule compared with the all too evident portion which falls on the near surroundings. Thus each nation will wish to clean up its air for the sake of its own citizenry.

Limiting the carbon dioxide content of the atmosphere presents knottier problems. The first necessary step is to decide on an acceptable upper limit for the atmospheric level. Once this has been done, it follows that combustion of fossil fuels must be paced so that when this level has been reached, only very limited burning of fossil fuels will continue beyond that time. But how is the allocation of a total acceptable combustion to each nation to be made? Clearly, it cannot be done on the basis of present rates of fossil fuel usage. Allowance must be made for the fact that most developing nations will be increasing their per capita usages of fossil fuels relatively more rapidly than the highly industrialized nations. Nor can the allocations be based on the known reserves. That is, it would not be feasible to assign to each nation an allotment based on the total estimated fossil fuel holdings of that nation and transferable to another nation with sale of the fuel. Aside from the probably insurmountable difficulties which would be encountered in reaching agreements on the size of each nation's proven reserves, there would be no simple way to take care of future discoveries. In addition, this criterion is ethically unacceptable.

An equitable means of allocating the permissible burning of fossil fuels would be to grant each nation a permit to burn a quantity of fuel based solely on its population. The permit would have nothing to do with ownership of the fuel itself. This would mean, for example, that the United States would receive a permit amounting to 7 percent of the world total of permissible future consumption. This, of course, is far less than the United States would like to have, since it is by far the largest consumer of fossil fuels. It might then negotiate with other nations to buy from them some portion of their permits. Presumably the prices paid for such permits would be substantial. The nations selling them could use the funds obtained in this way to improve technology, to build hydroelectric plants if appropriate, or to develop nuclear-fueled power plants. An unscrupulous ruler might well sell his nation's permit for personal gain, but this is already a practice with respect to mineral and other resources in some nations. The ideal arrangement would be one requiring that payment for burning permits be made in the form of nuclear fuel, construction of nuclear-fueled plants, or in other ways which would add to the selling nation's energy resources. The important point is that, in the majority of nations, in which the political leaders operate in the best interests of their countrymen as they see them, this scheme provides an opportunity for partially erasing the gap between the haves and have-nots. Citizens of the poorest nations are, if anything, more at the mercy of any undesirable climatic changes resulting from excessive carbon dioxide levels than those in the rich nations responsible for the excess. It does not seem at all logical that they should, as now, enjoy none of the pleasures and comforts which the burning of fossil fuels brings to affluent peoples, and yet share equally in the danger of possibly unhappy consequences.

In the operation of this plan it would be necessary for each nation to report regularly on the extent of fossil fuel consumption, and permits bought or sold. An international agency would be required to maintain records, monitor the atmospheric level of carbon dioxide and in general to police the use of fossil fuels.

A carbon dioxide level of about 450 parts per million would seem to be a reasonable limit to set initially. After reaching this level, and maintaining it for some time, detailed observations and data collection would permit a decision about the climatic effects. By that time, there may be little pressure for further fossil fuel consumption, because alternative methods of energy production will have necessarily come into widespread use in anticipation of a decision against further additions.

Implementation of the suggested plan would have interesting consequences for the development of nuclear power. In the United States and Europe, where the consumption of fossil fuels proceeds at a very high rate, there would be immediate interest in purchasing fossil fuel burning permits. On the other hand, the value of these permits would in all probability decline after a few decades, as the most highly developed nations close out fossil fueled power plants, replace the internal combustion engine, and substitute electricity for gas and oil in space heating. Early selling of permits would permit the developing nations to concentrate on nuclear-fueled power plants. Since much of the technology in developing nations is imported from the more highly developed, there would be little incentive after a few decades for instituting use of fossil fuels.

Of course these interchanges of technology would need to take place in the context of agreement of limitations in the total overall rate of energy consumption. Here again, it would be necessary to decide on a permissible total rate. The value of 6×10^{14} watts arrived at in Chapter 6 would certainly be a maximum possible rate; it could produce an increase of as much as 2°F in the average surface temperature, assuming that waste heat could be evenly distributed into the oceans. If methods for efficient transfer to the oceans are not worked out (the capability is not on hand at present), this rate of energy consumption would certainly produce excessive temperatures on land areas.

It would probably be most equitable to allot energy generation to nations on a strictly per capita basis, working from the population on an agreed-upon date, say the year 2000. The per capita allowance would be the maximum total rate agreed upon, say 5×10^{14} watts, divided by the world population at that time. The effect of an agreed-upon energy allotment would be to encourage population control. Population growth beyond the year 2000 would have no serious consequences for most nations because they would not be near the maximum allowed rate of energy dissipation, which would work out to be about 60,000 watts based on a population of 8 billion, correspondingly less for a larger population. It will require at least a century before most nations approach such a rate of energy generation. In the intervening time, it should prove possible to develop population control in such a way that the desired balance between population and per capita consumption can be attained and then maintained. Obviously, such a scheme cannot be put into effect until there is agreement on world-wide population control. Nations with high birth rates will not wish to settle for a total energy assessment which would require a much lower per capita energy quota when their limits have been arrived at in the distant future. It would perhaps be useful to establish the energy limit quotas gradually, beginning with an assignment of half of the total in the year 2000, then another quarter in the year 2025, and so on. During the intervening time other forces which are important in determining population growth will be operating. By 2050 the world will have solved the population problem or have fallen into more or less total disarray. Only the most highly advanced nations of the world will be approaching their energy limit at this time.

The need for world-wide agreements on control of atmospheric pollution, fossil fuel consumption and, ultimately, total rate of energy consumption is evident. The less prosperous peoples are entitled to a just share of what is obviously a limited resource, namely the capacity of the earth-atmosphere system to accept the products of industrial man. Mankind will be the better for a prompt recognition of the limits of allowable exploitation and establishment of an equitable distribution of the rights to use the earth, its waters and atmosphere. The technologically advanced nations could make no more important contribution to the future stability of the world than by assuming the initiative in formulating the required reforms. An international agency, perhaps under the auspices of the United Nations, should be established to monitor the planetary climate using all the most modern techniques. This agency should be vested with responsibility for monitoring on a nation-by-nation basis, use of fossil fuels, production of atmospheric pollutants of various kinds, and total energy consumption. Most of these monitoring activities could be carried out without undue interference by use of satellites and ground-based observations at appropriate locations, and on the basis of data furnished by individual nations. Even China, the most populous nation on earth, should find this an agreeable game to play. If she refused to participate, the plan could go on without her cooperation, since in any case she will be incapable for many years of exceeding her just share of any allocations.

When we turn from global considerations to the more localized issues created by energy generation in the United States, the time scale shifts. The problems of thermal pollution are with us in some localities now, and are becoming steadily more evident everywhere. During the summer of 1970 the city of New York went through several days with intermittent loss of power in sections of the city and a general reduction of 5 percent in the voltage during other critical periods. Much of the shortage of power can be traced to delays resulting from controversies regarding siting of new installations. Nuclear-fueled plants have come in for especially intense criticism because of their possibly adverse thermal effects, and because of concerns about their safety in terms of emissions of radioactivity.

The demand for power is intense; it penetrates the entire economic system and it will not be denied. At a recent symposium on power generation and environmental change it was pointed out that increased population accounts for only about 20 percent of the increased power consumption each year.[6] The remainder represents an increase in the per capita consumption of power. In response to the question of whether the society would, or could learn to, accommodate lower per capita power consumption levels, the speakers generally agreed that "society" would demand the power and that it would be forthcoming. At still another recent conference, on fuel technology, it was pointed out that there will inevitably be a chronic shortage of electrical power for the next several years.[7] New installations cannot be built rapidly enough to cope with the increasing demand. To add to the difficulties, there is a shortage of coal! Although the United States has enormous reserves, the economics of coal mining during the past few years have not been conducive to operation of mines at high capacity. A number of marginally successful mine operations have closed because of increased costs resulting from stricter safety and health regulations.

The ever increasing demand for more electric generating capacity will bring great pressures to bear on groups concerned with conserving the environment. Electric utility companies need only point to the environmentalists when complaints develop over brownouts and blackouts, and when shortage of power means the loss of a new industry or a new all-weather shopping complex. It is difficult in an atmosphere highly charged with conflict and laden with adversarial tactics to establish a reasonable long-range policy for the growth of electric power and other means of energy consumption. It seems evident, however, that a first step would be to set a reasonably firm upper limit on the total national rate of energy consumption of all kinds and then to establish a tentative upper limit on the total electric generating capacity. Although these limits might appear to be of little relevance in connection with questions relating to present controversies over specific sitings, they could be of immense importance in setting a tone which is now entirely absent from the posture of the electric utility industry. No one in that industry is willing to estimate the limits on its growth. The industry has in fact done everything possible, in the venerated American tradition, to encourage the growth of demand for its product. It would ill-behoove the executive officers of an investor-owned corporation to do otherwise. Of course, there is a limit; who would argue that there is not? It's just that no one wants to say what it is. Nor will anyone in the industry confirm that, in terms of present technology, the rate of growth in demand for electric power is too high. It is apparent, however, that the present growth rate of the electric power industry must be slowed to avoid serious damage to the environment resulting from hasty and ill-conceived construction of power plants.

To see that this is so we need only consider what must happen if the present growth rate continues. Electric power consumption is doubling every ten years; in the absence of control measures, it could very well continue to double at this rate well into the next century. In Chapter 7 we saw that the addition of about 10,000 megawatts of new nuclear power capacity to the shores of Lake Michigan could result in an increase in lake temperature by as much as 0.1°F. Since this 10,000 megawatts would be in addition to the 6,000 megawatts of fossil-fueled plants now using the lake for cooling, the total temperature increase above the natural temperature would be as much as 0.15°F. We can expect this to be the situation in about 1985. But by 1995 there will need to be twice this much capacity on the lake, so the lake temperature might be as much as 0.3°F above normal; by 2005, twice as much again, to produce up to a 0.6°F rise in temperature; by 2015 another doubling, and up to 1.2°F; and so forth. With less than ideal mixing, as assumed in making the temperature estimates, areas of the lake bordering on the power stations might be maintained at temperatures much in excess of the calculated rises.[8]

To anyone who reasonably suggests that of course the electric generating capacity is not going to keep doubling at this rate, I can only point out that this is the rate at which electric power consumption has been growing for the past three decades; this is the rate at which it is growing today; this is the rate at which the power industry itself anticipates it will grow for at least the next thirty years. Perhaps after a time it won't continue to increase so rapidly. It may require twenty years to double instead of ten, so that it will be as late as perhaps 2030 before Lake Michigan is being heated 1.2°F above the natural level. Regardless of the precise rate at which it occurs, if present policies and practices are followed in the future, the lake will be heated considerably. It has been noted that the present temperature of Lake Michigan may actually be about 2°F lower than it was 100 years ago. Thus, all of the anticipated heating would in effect merely return the lake to a previous condition. But the condition of the lake in respects other than temperature is far different now than it was 100 years ago. The temperature of a lake which is not rich in nutrients, and which thus does not support a very substantial level of biological activity, is not so critical in affecting the rate of eutrophication of the lake. During the past century Lake Michigan has been converted from an oligotrophic lake (that is, one low in nutrient content) to a relatively rich body of water, eutrophying at an undesirably high rate.[9] Until the continuing pollution of the lake with phosphates, nitrates and other nutrients in the effluents of municipal sewage treatment plants and industrial installations is halted, and until the lake is cleansed of the high levels of accumulated nutrients, *any* temperature rise in the lake is detrimental. Aside from the deleterious effects of increased temperature on lake ecology, the consumptive use of water through evaporative loss will be increasing along with the total electric generating capacity located on the lake. This consumptive loss would be even more serious if cooling towers were employed.

We can conclude from these considerations that, in terms

of the technology now employed in the use of lake water, the maximum electric generating capacity which might safely be located on Lake Michigan is probably that anticipated for about 1985, when there might be about 10,000 megawatts of new nuclear-fueled generating capacity installed. Now is the time to start worrying about where we go from there.

If the situation with respect to installations on Lake Michigan looks troublesome for a decade or so from now, the more usual case, represented by Cayuga Lake, is even more immediately critical. The electric utility plan to employ an open, single-pass cooling system is subject to criticism on the grounds that it will promote an increased rate of eutrophication, in part by causing a substantial increase in average lake temperature during the growing season. Furthermore, the consumptive use of water might be excessive in terms of the natural inputs to the lake. But if the original installation, now scheduled for completion after 1973, is properly subject to criticism, what of the prospects that the capacity will need to be doubled during the 1980s and quadrupled during the 1990s? The dates may be changed by alteration in the rate of growth of the New York Electric and Gas Corporation, but a few years more or less are not the issue. The essential point is that generating capacity in addition to the 830-megawatt unit now under construction will exacerbate what will already be a damaging situation. The fact that the utility has made a strong bid for the Cayuga Lake installation is in itself indication that alternative locations for large power plants are not plentiful. If the demand for power is to grow unchecked, the cooling water resources must come from somewhere. It will not be long before all the larger fresh water lakes, rivers and streams are dedicated to the business of cooling power plants. As an example, in testimony before the Senate Sub-committee on Air and Water Pollution it has been estimated that there are about 3,700 miles of river in the United States which might have sufficient flow rates to accept huge 2,000-megawatt plants along the banks at ten-mile intervals, each discharging water at up to 105°F into the rivers.[10] This massive array of power plants strung out along the Mississippi, Ohio, Missouri, and other rivers would generate 740,000 megawatts of power. But even this alarming concentration of electric generating capacity would provide only 70 percent of the estimated national requirement for the year 2000!

It is evident that a slowing in the growth rate of the electric power industry is of itself an inadequate response to the impending problems. Slowing the growth rate merely delays the inevitable. As the United States moves beyond the present per capita overall energy consumption of about 10,000 watts, the increase must come mainly from increased electric power consumption. This fact is evident in the energy forcast depicted in Figure 6-1, page 69. But if this growth is to occur in an orderly manner, the management of waste heat must be vastly improved over the present practices. The long range solution would seem to lie in locating a large fraction of the new capacity on or in the oceans.[11] Transmission of power to inland sites by overhead transmission lines should be forbidden as an utterly unacceptable disfigurement of the landscape. A vast research effort on the part of the electric utility industry, which has an enormous gross sales and spends almost none of it on fundamental research, would probably lead to rapid development of acceptable, underground transmission lines. Unless and until significant new developments can be brought to the point of successful application, the growth of the electric utility industry should be tightly controlled. The alternative to controlled growth and a vastly improved technology is a grossly uneven distribution of heat generation, with drastic deterioration in regional climates and spoilage of public waters.

There are many ways in which the present growth of electric power consumption could be moderated. Electricity is far too inexpensive, considering the stresses placed on the environment during its production. It would be appropriate, for example, to charge utilities realistically for gross use, and most especially for consumptive use, of fresh water from public waterways. It would be appropriate to require utilities to establish a different rate structure, so that massive users of electrical power such as the aluminum industry would not in effect receive a subsidy from the average homeowner for the vast amounts of power consumed in producing aluminum from raw ore. Recycling of aluminum requires only a small fraction of the power required to produce the metal from the raw ore. Since aluminum production consumes about 10 percent of all industrial power use, economic inducements in just this one area would serve to moderate growth in electricity consumption, as well as ameliorate a serious problem in disposing of waste aluminum. Pricing electricity more realistically—that is, in keeping with the real costs to the society of its generation—would encourage research in more effective methods of lighting. It borders on the ridiculous that the most advanced society on earth still relies to a large extent on the incandescent light bulb for generation of visible light. The incandescent light bulb is simply a black body radiator (see Chapter 3); only a fraction of the energy put out by the bulb falls in the visible region of the spectrum. The remainder is wasted. Fluorescent lighting is only one of many alternatives which might be developed if the incentives were there.

The electric industry has fostered the notion that electric home heating is efficient and clean. It is of course true that in the home itself all of the electricity consumed goes directly into heat in the home, whereas some of the heat produced in an oil or gas furnace goes up the chimney. But it must be remembered that a great deal of waste heat is released when the electricity is generated in the first place. In terms of overall efficiency, electric heating of homes is the most wasteful of all commonly employed methods of central home heating.

Many scientific and engineering tricks might be employed to increase the efficiency of energy use in a variety of contexts. For the most part these have not gained widespread use and have not even been properly researched because the economic incentives have been absent.

If the general public is to be sufficiently aroused by the dangers involved in the unchecked growth of the electric utility industry, it must be provided the opportunity to understand the issues involved. The Federal Power Commission and agencies of the government charged with responsibility for preservation of the environment must spell out as clearly as possible the costs of electric power, and how these costs might be affected by changes in rate structure and other controls on the growth of the industry. The federal government should encourage development of higher efficiency generation, transmission and use of electric energy. Given that the increase in generating capacity must be limited, steps should be taken to prevent untrammeled growth in demand. It is sheer irresponsibility for New York and Chicago to encourage the construction of huge new skyscrapers with their extraordinarily high power demands when these cities are already chronically short of power.[12]

For the United States and a few other highly industrialized nations the limited capacity of the environment to cope with man's output is becoming increasingly clear. Those who insist upon thinking in terms of the open, limitless economics which Boulding terms "cowboy economics" may find the notion of a closed, limited system disconcerting. But instead of diminishing our capacity for innovation, the new challenges created by success should occasion new enthusiasms and renewed activity. It will require a high order of inventiveness to make effective use of what we now recognize as an ultimately limited rate of energy consumption. It will require a commensurate level of inventiveness to produce a social, political and economic order in which all men share in the good things of life on our home, the earth.

notes

Chapter 1

1. While ozone is, from our point of view, a friendly substance at an altitude of 20 miles, it is very unpleasant at close range. It is one of the leading villains in the smog scene. During a heavy siege of smog in a city such as Los Angeles, the ozone level may rise as high as 4 parts in ten million, at which level it produces a variety of unpleasant symptoms and unhealthful results.

Chapter 2

1. Satellite Observations of the Earth's Radiation Budget, T. H. Vonder Haar and V. E. Suomi, *Science*, 1969, vol. 169, 667.

2. The Stefan-Boltzmann law relates the total emissive power to the fourth power of the absolute temperature: $E = \sigma T^4$, where σ is a proportionality constant, and T is the temperature in ° Kelvin.

Chapter 3

1. (a) Satellite Observations of the Earth's Radiation Budget, T. H. Vonder Haar and V. E. Suomi, *Science*, 1969, vol. 169, p. 667.

(b) The Radiation Balance of the Planet Earth from Radiation Measurements of the Satellite Nimbus II, W. R. Bandeen, *Journal of Appl. Meteorology*, 1970, vol. 9, p. 215.

2. *General Meteorology*, 3rd Edition, H. B. Byers, McGraw-Hill Book Company, New York, 1959, Chapter 3.

3. Thermal Equilibrium of the Atmosphere with a Given Distribution of Relative Humidity, S. Manabe and R. T. Wetherald, *Journal of Atmospheric Sciences*, 1967, vol. 24, p. 241.

4. Atmospheric Temperature: Successful Test of Remote Probing, D. Q. Wark and D. T. Hilleary, *Science*, 1969, vol. 165, p. 1256.

Chapter 4

1. (a) The Influence of Infrared Absorptive Molecules on the Climate, G. N. Plass, *Annals of the New York Academy of Sciences*, 1961, vol. 95, p. 61.
(b) Carbon Dioxide and Climate, G. N. Plass, *Scientific American*, 1959, vol. 201, p. 41.

2. Thermal Equilibrium of the Atmosphere with a Given Distribution of Relative Humidity, S. Manabe and R. T. Wetherald, *Journal of Atmospheric Sciences*, 1967, vol. 24, p. 241.

3. Relative humidity gives the amount of water vapor present as a percent of what is required to saturate the atmosphere under the particular conditions. A 50 percent relative humidity at –50°F, for example, means far less water vapor per unit volume than a 50 percent relative humidity at +75°F.

4. Climate Calculations with a Combined Ocean Atmosphere Model, S. Manabe and K. Bryan, *Journal of Atmospheric Sciences*, 1969, vol. 26, p. 786.

5. *Resources and Man*, W. H. Freeman and Company, San Francisco, 1969, Chapter 8 by M. King Hubbert.

6. *Power Engineering*, December 1968, p. 50.

7. Equilibration of Carbon Dioxide with Sea Water: Possible Enzymatic Control of the Rate. R. Berger and W. F. Libby, *Science*, 1969, vol. 164, p. 1395.

8. Radioisotopes and Large Scale Oceanic Mixing, W. S. Broecker, in *The Sea*, M. N. Hill, editor, Interscience, New York, 1963, vol. 2, p. 88.

9. pCO_2 in Sea Water, and Its Effect on the Movement of

CO_2 in Nature, J. Kanwisher, *Tellus*, 1960, vol. 12, p. 213.

10. Degree of Saturation of $CaCO_3$ in the Oceans, Y-H Li, T. Takahashi and W. S. Broecker, *Journal of Geophysical Research*, 1969, vol. 75, p. 5507.

11. There is some evidence that this simple "two-box" model of the ocean is an oversimplification and that more sophisticated considerations are involved: Variations in the Carbon Dioxide Content of the Atmosphere in the Northern Hemisphere, B. Bolin and W. Bischof, *Tellus*, 1970, vol. 22, p. 431. The simple model does, however, lead to predictions which are roughly correct!

12. (a) Large Scale Atmospheric Mixing as Deduced from the Seasonal and Meridional Variations of Carbon Dioxide, B. Bolin and C. D. Keeling, *Journal of Geophysical Research*, 1963, vol. 68, p. 3899.
(b) The Concentration of Atmospheric Carbon Dioxide in Hawaii, J. C. Pales and C. D. Keeling, *Journal of Geophysical Research*, 1965, vol. 70, p. 6053.

13. *Climate Through the Ages*, C. E. P. Brooks, Dover Publications, Inc., New York, 1970, Chapter 8.

14. Controlling the Planet's Climate, J. O. Fletcher, *Impact of Science on Society*, 1969, vol. 19, p. 151.

15. *Weather and Climate Modification, Problems and Prospects.* Volume II Research and Development. Publication No. 1350, National Academy of Sciences-National Research Council, Washington, D. C., 1966, p. 83.

16. Possible Changes in Atmospheric Carbon Dioxide Due to Changes in the Properties of the Sea, E. Eriksson, *Journal of Geophysical Research*, 1963, vol. 68, p. 3871.

Chapter 5

1. *Weather and Climate Modification, Problems and Prospects.* Volume II Research and Development. Publication No. 1350, National Academy of Sciences-National Research Council, Washington, D. C., 1966.

2. Radiation Balance of the Earth as a Factor in Climate Change, H. Wexler, Chapter 5 in *Climate Change: Evidence, Causes and Effects*. H. Shapley, Ed., Harvard University Press, Cambridge, Massachusetts, 1953.

3. A Reconciliation of Several Theories of Climatic

Change, Reid A. Bryson, *Weatherwise*, April, 1968, p. 56.

4. *Air Quality Criteria for Particulate Matter*. AP-49 National Air Pollution Control Administration, Washington, D. C., January, 1969, Chapter 2.

5. *Ibid.*, Chapter 11.

6. Our SST and Its Economics, G. M. Swihart, *Astronautics and Aeronautics*, April, 1970, p. 32.

7. *SST and Sonic Boom Handbook*, W. A. Shurcliff, Ballantine Books, 36 W. 20th St., New York, N. Y., 10013, 1970. $0.95. See Appendix 4.

8. Water Vapour Pollution in the Stratosphere by the Supersonic Transporters? R. E. Newell, *Nature*, 1970, vol. 226, p. 70.

9. Estimates of water vapor level changes in the stratosphere are very tentative because they depend, among other things, on assumptions about latitudinal as well as altitude variations. See also *Man's Impact on the Global Environment*, MIT Press, Cambridge, Mass., 1970, pp. 64–74; 100–107.

10. The New York Times, May 28, 1970.

11. Insolation in Relation to Cloud Type, B. Haurwitz, *Journal of Meteorology*, 1948, vol. 5, p. 110.

12. Thermal Equilibrium of the Atmosphere with a Given Distribution of Relative Humidity, S. Manabe and R. T. Wetherald, *Journal of Atmospheric Sciences*, 1967, vol. 24, p. 241.

Chapter 6

1. *Resources and Man*, W. H. Freeman and Company, San Francisco, 1969, Chapter 8 by M. King Hubbert.

2. *Power Engineering*, December, 1968, p. 50.

3. *The Careless Atom*, Sheldon Novick, Dell Publishing Company, New York, 1969.

4. Limits to the Use of Energy, A. M. Weinberg and R. P. Hammond, *American Scientist*, 1970, vol. 58, p. 412.

5. Thermal Equilibrium of the Atmosphere with a Given Distribution of Relative Humidity, S. Manabe and R. T.

Wetherald, *Journal of Atmospheric Sciences,* 1967, vol. 24, p. 241.

6. One-half percent of the *total* insolation represents about 0.7 percent of the averaged insolation at the surface after correcting for atmospheric absorption. This would probably cause an average temperature increase on the order of 1 to 1.5°F, assuming that heat could be widely distributed.

7. *The Nuclear Years:* The Arms Race and Arms Control, C. M. Roberts, Praeger, 1970.

8. A National Estimate of Public and Industrial Heat Rejection Requirements by Decades through the Year 2000 AD, R. T. Jaske, J. F. Fletcher and R. R. Wise, Batelle Memorial Institute, Pacific Northwest Laboratory, February, 1970. Presented at the 67th National Meeting of the American Institute of Chemical Engineers, Atlanta, Georgia, February 17, 1970.

9. Man-made Climatic Changes, H. E. Landsberg, *Science,* 1970, vol. 170, p. 1265.

Chapter 7

1. *Problems in disposal of waste heat from steam-electric plants,* Federal Power Commission-Bureau of Power, 1969.

2. *Thermal Pollution:* Status of the Art, F. L. Parker and P. A. Krenkel, Vanderbilt University, December, 1969.

3. *Thermal pollution*-1968 Hearings before the Sub-committee on Air and Water Pollution, Committee on Public Works, U. S. Senate, 1968, Parts 1–4.

Chapter 8

1. A New River, *Environment,* Jan.-Feb., 1970.

2. This estimate can be arrived at by choosing appropriate values for the parameters in equations developed in the literature, e.g., "Water Loss Investigations: Lake Hefner Studies." Technical Report, U. S. Geological Survey, Professional Paper 269. Alternatively, it can be based upon actual energy budget data for a typical lake or reservoir, e.g., the results of R. N. Bergstrom depicted on page

VI-5, Reference 2 of Chapter 7. The most important variable in considering the applicability of any calculation to a particular water body is the evaporative loss rate. Studies of the evaporation from Lake Ontario and Lake Superior, as described in the following reference[3], indicate that an annual average of about 2×10^5 BTU/acre–hr/°F for the total net change of heat transfer as a function of temperature is reasonable. About half this, on the average, is due to evaporative heat transfer.

It is noteworthy that the estimate of temperature change made by J. P. Longtin of the Federal Water Pollution Control Administration (see reference 4 below) is essentially the same as the estimate I have made. Longtin calculates a temperature increase in the surface of the *entire* lake of about 0.04°F for a heat rejection of 23,000 megawatts, assuming perfect mixing.

3. (a) Evaporation from Deep Water Lakes, F. Miller, *Water Resources Research,* 1967, vol. 3, p. 181.
(b) Stochastic Aspects of Lake Ontario Evaporation, S. L. Yu and W. Brutsaert, *Water Resources Research,* 1969, vol. 5, p. 1256.

4. Proceedings of the Conference on Pollution of Lake Michigan, second session, Feb., 1969, Volume 2, p. 671. U. S. Department of The Interior, Federal Water Pollution Control Administration.

5. *The Ithaca Journal,* Ithaca, New York, December 13, 1968.

6. Pollution Problems, Resource Policy and the Scientist, A. W. Eipper, *Science,* vol. 169, p. 11 (1970).

7. *The Ithaca Journal,* Ithaca, New York, November 28, 1969.

8. This is a reasonable assumption, based upon the data available for nearby Lake Ontario (reference 3 above).

9. Most of the details regarding The Vermont Yankee installation are to be found in Parts 1 and 2 of Reference 3, Chapter 7.

10. Possible Environmental Effects of the Cooling Tower Installation for The Vermont Yankee Nuclear Power Plant, J. C. Harding, Thayer School of Engineering, Dartmouth College, June, 1970. The author is appreciative of this and other information provided by Professor A. O. Converse of Dartmouth College.

11. Changing Patterns in Energy Transport, O. B. Falls, Jr., *Power Engineering,* May, 1969, p. 28.

12. *Considerations Affecting Steam Power Plant Site Selection,* Office of Science and Technology, Energy Policy Staff, December, 1968.

13. *Thermal Pollution:* Status of the Art, F. L. Parker and P. A. Krenkel, Vanderbilt University, December, 1969.

14. On Possible Constructive Uses of Thermal Additions to Estuaries, J. A. Milhursky, *Bioscience,* 1967, vol. 17, p. 698.

15. Heated Effluent: An Asset to Agriculture? L. B. Bradley, *Power Engineering,* March, 1969.

Chapter 9

1. Man-made Climatic Changes, H. E. Landsberg, *Science,* 1970, vol. 170, p. 1265.

2. *Descriptive Palaeoclimatology,* A. E. M. Nairn, Editor, Interscience Publishers, New York, 1961.

3. Recent results based on oxygen isotope studies of the Greenland ice sheet reveal temperature variations which can be related to evidence for concurrent temperature variations elsewhere on the planet: One Thousand Centuries of Climatic Record from Camp Century on the Greenland Ice Sheet, W. Dansgaard, S. J. Johnsen, J. Möller and C. C. Langway, Jr., *Science,* 1969, vol. 166, p. 377. Climatic Oscillations 1200–2000 AD, S. J. Johnsen, W. Dansgaard, H. B. Clausen and C. C. Langway, Jr., *Nature,* 1970, vol. 227, p. 482.

4. The Economics of the Coming Spaceship Earth, K. E. Boulding, *The Environmental Handbook,* Garrett de Bell, Editor, Ballantine Books, New York, 1970.

5. What We Must Do, John Platt, *Science,* 1969, vol. 166, p. 1115.

6. Power Generation: The Next 30 Years, R. W. Holcomb, *Science,* 1970, vol. 167, p. 159.

7. Solutions Unclear for Power Shortage, *Chemical and Engineering News,* June 15, 1970, p. 16.

8. It is of interest that in testimony before The Conference on Pollution of Lake Michigan and Its Tributary Ba-

sin (Reference 4 of Chapter 8), F. W. Kittrell of The Federal Water Pollution Control Administration estimates on the basis of "conservative calculations" that the temperature increase in the lake might be 2.0°F by 2023.

9. Eutrophication of The Saint Lawrence Great Lakes, A. M. Beeton, *Limnology and Oceanography,* 1965, vol. 10, p. 240.

10. *Thermal pollution*-1968 Hearings before the Sub-committee on Air and Water Pollution, Committee on Public Works, U. S. Senate, 1968.

11. This prospect has already been discussed to a limited extent; see, An Oceanographic and Suboceanic Probe: Island Power Stations, J. R. McFarland, in *Pacem in Maribus,* an occasional paper of The Center for the Study of Democratic Institutions, Santa Barbara, California, 1970.

12. Power, Pollution and the Imperiled Environment, G. D. Friedlander, *IEEE Spectrum,* 1970, vol. 7, no. 11, p. 40. See particularly page 43.

index